Applied Plant Geography

Applied Plant Geography

Dr. Pramod Pagare
Head Department of Geography
Govt. M.G.M.P.G. College,
Itarsi (M.P.)

Oxford Book Compan

Published By

Oxford Book Company

Regd. Off. "Bony Residency", Gate No. 2
Opp. Tilak Public School, Vishweswariya Nagar,
Gopalpura Road, Jaipur - 302018, India
Ph: 0141 - 2761280, Telefax: 2761381
email - oxfordbookjpr@gmail.com
website - oxfordbookcompany.com

Delhi Office: 102, 1st Floor, Satyam House,
4327/3, Ansari Road, Darya Ganj,
New Delhi - 110002, India
Ph: 011-45652440

Distributed by
KEYA INTERNATIONAL PRESS LLP
E-mail: keyainternationalpressllp@gmail.com

Published in India, 2026

ISBN : 9789380179100

Typeset by Shivangi Computers, Jaipur

Printed and bound in India by Replika Press Pvt. Ltd.

Preface

The plants may well be considered as the wealth of environment and nature's banks. Plants (flora) is a peculiar organism of unlimited kindness and benevolence. Since time immemorial plants have played significant role in shaping the environmental quality and maintaining ecological balance to sustain life. They also determine the site productivity, nutrient cycling and energy potential. In this way plants exert multifarious influence on mode of life on earth.

The relationship between mankind and plants has a complex and ambivalent history. They not only determine the natural conditions of the man, but also provide nourishment and protective substance for good health of human being. God was kind enough to bestow the mankind with diversified resources of flora, of course he has left the task of recognizing their curative values to man. Thus, this calls for investigating the potentiality of applied values of plant resources in the region.

Forest in India as elsewhere in the world are significant to great extent both from the ecological and development points of view. In Central India, forest ecology is one of the dominating features of regional personality. The rich forest covers of Satpuras with distinct flora provide potential fields to the biogeographers for investigation. But it is unfortunate that the biogeographers have so far shown little interest in the studies of forest resources of the Central India. It is a sorry state of affairs that in spite of all this no serious and systematic work regarding the forests of Satpura has been carried out either from ecological or geographical point of view.

The present study will include environment and spatial distribution of valuable plant species and is a humble attempt to analyze the environmental condition of plants of applied values and to suggest conservative measures of such plant species. This study will also focus on the forest ecology of the Satpura forest and enrich the knowledge in the domain of plant geography, a subfield of the Biogeography. It is hoped that this book will encourage future researchers, common readers and planners who are engaged in such studies of plant resources.

My deep felt thanks are to the conservator of forest, Betul and his staff members for helping me in the field survey, without whose help it would have been a tough task. I have profusely benefited from the discussions with several herbalists, botanists, vaidyas and villagers. I would like to express my thanks to all of them. This book would not have been possible without the love and perseverance of my wife Jaya Pagare, parents and family members, my indebtedness to them is beyond words.

Dr. Pramod Pagare

Contents

1
Introduction

INTRODUCTION

Geography is an important member of the environmental sciences. In fact it is geography which presents the integrated and wholistic picture of the environment. Among the geographical traditions, ecological tradition has been a significant approach of study which interrelates human and environmental variables and also interprets their links (Haggett et al., 1977, 5). The recent association of geography and environment under Bio- geography will provide a strong platform and strategy for protecting, conserving and transforming the environment and its resources to promote integrated welfare of a society .

Biogeography is primarily concerned with interpreting the living endowment at a point on the earth as a qualitative indicator for the corresponding space e.g. by area system analysis (Miller, 1976). The study of historical biogeography, the fascinating ecology and floral as well as faunal history of a land is really the subject of biography and evolution (Dansereau 1957). Similarly the study deals with the effect of environmental factors on the functional aspects of organisms like plants and animals are the subject matter of

Physiological ecology of Bio-geography (Barry, 1977). The Ecological geography dealing only with plant communities and vegetation as called phytogeography.

Phytogeography is a Greek word which means: "phyto-plants, geo-earth" and "graphein to write or to describe". Hence, this branch of Biogeography deals with correlation among the plant of the earth and their geography. Hardy (1913, 16) also expressed similar views that the study of the distribution of different types of vegetation and the way in which these types and their distribution are related to the other geographical condition is called Plant geography or Phytogeography. The classical approach of phytogeographical studies has been towards enumeration of the taxa of a region and on the basis of broad floristic differences in ecological regions has been recognized. Thus, phytogeography has two major approachas of study (A) descriptive or static phytogeography dealing with description of flora or vegetation of different region and (B) interpretive or dynamic phytogeography dealing with interpretations of causes of plant distribution (Ambasht 1976).

The Applied Phytogeography is an interdisciplinary approach of the above mentioned branch (Phytogeography) which deals with the geographical environment of the plant species which have applied aspect for the welfare of human beings as well as for domestic animals. Thus, it is one of the prime duties of an applied phytogeographer to investigate the potentiality of applied values of existing plant resources of a particular region. Further in this context, he must be aware of the traditional wisdom of the native inhabitants who are repository of accumulated experience and knowledge of indigenous vegetation to which they transfer from generation to generation (Charan 1991 and Maheshwari 1990, 64).

It is evident from the recent survey conducted by the Indian Council of Social Science Research (ICSSR) that there is an imbalance contribution of geographers in the field of phytogeography, especially the studies so far published have not talked about the plant kingdom as a resource and the environmental impacts on the availability and other physical and biological aspects have also not been touched upon. Looking to the need of the day, the time has come, when a critical and extensive study have attracted the attention of the geographer.

In view of above, the study of the distribution of different types of medicinal plants and the way in which these types and distribution are related to the other environmental conditions, are ought to be studied yet.

The scientists belonging to the natural science like botany, chemistry and pharmacy have analysed and classified the plant kingdom but the nature of their distribution and their relationship with environmental factors and forces have not been highlighted with the view of assesing and to determine their potentialities and the problems associated with the sufficiency and availability of such important plant resources. Such studies are very much within the orbit of geographical studies. (Hartshorne, 1939) Hence, the study of plants as a resource has become more important in the context of present global trade scenario. In such time it is highly needed that more folklore of our indigenous knowledge of plant resources should be explored and the availability of such plants should also be surveyed in each quarter of the state. The environmental system which is promising and favourable to the growth of such plant species to be determined in order to propagate, preserve and to plan for future.

Unfortunately, the major management projects have not been based on research and development studies aimed at systematically bridging these gaps in knowledge (Kunkle and Thames, 1977). It is therefore, essential to investigate the environmental conditions and the intensity of the plant resources in different parts of the world which will help in both ways for understanding and in maintaining equilibrium between human population and the plant kingdom as a resource.

Evolution of Plant Geography

The history of phytogeography is as old as the ideas of history of nature. A brief account of the growth of plant geography as a part of systematic geography may held placing the following study in a proper historical context. According to German Geographer Alexander Ven Humboldt phytogeography started from Linnaeus's work in 1753. Humboldt (1807) specialized in the native flora and the crop plants of Switzerland and France. He published various accounts of his travel and these works present his phytogeographical observations.

The first new geographer Sir Joseph Dalton Hooker made several major contributions to plant taxonomy by an expedition into the higher regions of Himalayas in 1846. Hooker along with T. Thomson published their work from London in 1855 entitled FLORA INDIA and also in 1872-79 under the title The Flora of British India.

In the late nineteenth Century emerged a German school of plant geographers. It included men such as O. Drude, E. Geoze, A.H.R. Griseback and A. Henson and each of them wrote a book about plant distribution on a world-wide scale. Among the Geographers belonging to the school was Heinrich Gustav, Adelf Engter who published a series of work during 1872-1882 pertaining to the plant geography.

In the beginning of twentieth century the botanists Schimper (1903-1904). Thode (1925), Salisbary (1916) also promoted plant geography through their work. The progress of botanical collection had promoted refinement of Engler's plant geographic units.

During the same period V.V. Alekhin in Russia, C.P. Black and A.C. Atwood in the United states, G.S. Boulger and M.E. Hardy in England, Leon Croizat in Netherland, H. Gaussen in France and A. Hayck in Germany have written about plant geography of the world.

The subject of phytogeography was also studied in various aspects by some authorities from time to time. Raunkiaer (1907) studied the plant life and published his work in his book. The life form of plants and their bearing on geography in 1907 and The life form of plants and statistical geography in 1934.

Another pioneer in this field is H.G. Champion whose contributions come mostly in the shape of research articles and papers. In 1920-36 he contributed research articles on environmental approach to the study of forests of India.

A modern plant geography scheme was firstly presented by Ronald D. Good. His work on Geography of the flowering plants was published in 1947. In 1957 Dancereau divided the globe into six major phytogeographical kingdoms with 37 floristic provinces.

After Clark (1898), Hooker (1906), Chatterji (1939), Razi (1955), Champion and Seth (1968) gave classification of vegetational type in India. Chatterji divided Indian continent into eight main floristic regions on the basis of natural rainfall.

During the second half of the twentieth century all the major contributions were from botanist and specialists. Among them the notable work are that of Kashyap (1925-32), Kambush (1924), Vishwanathan (1931-32), Khan (1935), Agarkar (1942), Bose (1948-49), Luthra (1953), Karnik (1954-56), Das and Sarup (1955) and Dansereau (1957).

For Indian subcontinent, bioclimatology which was a branch of biogeography appreciably studied by Bagnouls and Gausen (1957) and Bharucha, Mehar Homji (1965).

In 1960's Puri (1960), Dev (1960), Rao (1966-69), Sharma (1967), Verma (1967), have contributed in this field. Puri has contributed to the Indian Forest Ecology. He emphasized upon the vegetational characteristics with the environmental adaptability in botanical regions of India.

During the period of 1970's and onward the distribution of vegetation in spatial context with ecological perspectives has been highlighted by Cain (1971), Watts (1971), and Mani (1974).

During the later half of twentieth century a new branch called Applied Phytogeogrophy developed. In 1971 Shri Kamal Sharma emphasized on conservation of forest resources.

By realizing the best field of subject and its development in the modern age, Miller (1976) studied about vegetation geography and Geo-medicine as a branch of biogeography.

Several other phytogeographers have made valuable contributions in this field of phytogeography and their contribution come mostly in the shape of research articles and papers. In this field S. Edward Ayensci (1979) gave his view in his article Plants of

PROFILE OF THE STUDY AREA

The Betul Plateau, the mid-southern most Madhya Pradesh has been selected for the present study which covers the entire area of the Betul District lying between 21°22' north to 22°24' north latitudes and 77°4' east to 78°33' east longitudes, it forms a compact shape, almost a square with slight projections on the east and the west direction. The plateau runs almost along the Barar plains towards the south and the Narmada valley towards the north. It is bounded by the Hoshangabad district in the north, Amraoti district of Maharashtra in the south, the Chhindwara district in the east and in the west by the district of Hoshangabad,east Nimar and Amraoti (Plate 1.1).

The total geographical area of the study region is 10,059 square kilometres which is 3.26 per cent of the total area of Madhya Pradesh. The plateau stands at a height of about 700 metres above the sea level. Its terrain condition is characterized by a series of large and small plateau and hills intersected with different rivers, basins and valleys.

The population of the study area is 13,94,421 in 2001. Of the total population, about 81 per cent lived in rural areas, suggesting the predominance of rural character of the people. Corresponingly, the level of urbanization is low. About 19 per cent population lived in urban areas. About 27.95 per cent population is literate and sex ratio is 966 females per thousand males in the plateau. The total working population is 48.4 per cent of the total population.

From the socio and cultural point of view the study area constitutes a significant proportion of tribal inhabitants. It has 38.3 per cent scheduled tribes of the total population of which Gond and Korku are the main primitive human societies. They live in perfect harmony with nature for their survival. Gonds are numerically the largest tribal group of the study area; inhabit the maximum part of forest area while the Asir hill is the real home of Korkus tribes. The circumstances under which these people live object poverty, disease and hunger combined with curiosity towards their closest neighbour. They utilize many species of forest flora for food, medicine and other purposes.

PLATE 1.1

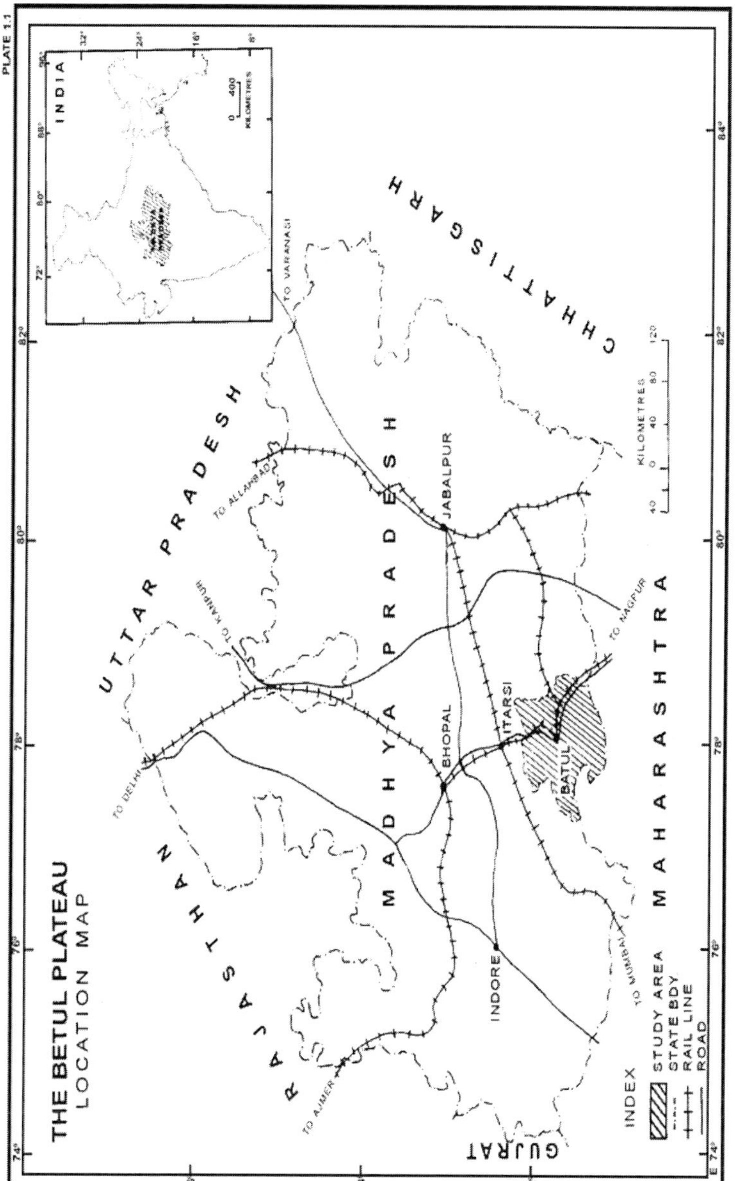
THE BETUL PLATEAU
LOCATION MAP
INDIA
UTTAR PRADESH
MADHYA PRADESH
RAJASTHAN
CHHATTISGARH
MAHARASHTRA
GUJRAT
JABALPUR
BHOPAL
ITARSI
BATUL
INDORE
TO DELHI
TO KANPUR
TO ALLAHBAD
TO VARANASI
TO NAGPUR
TO MUMBAI
TO AJMER
KILOMETRES
INDEX
STUDY AREA
STATE BDY
RAIL LINE
ROAD

In rural and remote areas of the study region people still depend upon traditional health care system due to insufficient medical facilities. It is one of the ancient care systems which is even now widely practised in tribal area *(Choubey, 1994, 73)*. People of this region have the knowledge of indigenous vegetation and they use different parts of plants for treatment of ailments. In this way this ancient system of medicine is deep rooted in the study area which leads to develop the traditional health care system.

In the study area, forests are one of the major forms of natural land scape and are the most important natural resources. About 40.1 per cent of the land area has been occupied by forests which is greater than the total forest cover of Madhya pradesh i.e. 20.75 per cent *(Central Forestry Commission, 1980, NRSA)*. The percapita forest area in the Betul plateau is 0.34 hectares whereas in India it is estimated only 0.06 hectares *(Garg and Bandhu, 1986, 151)*.

Now the study area falls under the Betul circle of forests of Madhya Pradesh which consists three forest divisions i.e. the north, the south and the west (Plate 1.2). The most common forest type is tropical deciduous that constitutes one of the prized asset of man. Though forests have economic significance in the study area as they are the natural sources of timber wood, fuel wood and valuable medicinal plants by which a considerable revenue is collected but there has been a scale erosion of the forest resources over the last three decades creating a serious imbalance in the ecosystem.

DATA BASE AND METHODOLOGY

The empirical approach has been one of the basic considerations to generate the primary data for the present study. The investigator visited the accessible part of the area under study to collect the required data regarding vicinal location of different plant species in this region of Madhya Pradesh. Some statistical tools are used in the work for detailed study pertaining to the distribution, population, and density of the plant species. On the basis of terrain type, three methods have been applied for field survey of plant species in the study area. Besides, various sampling techniques have also been used in order to get accurate quantitative information.

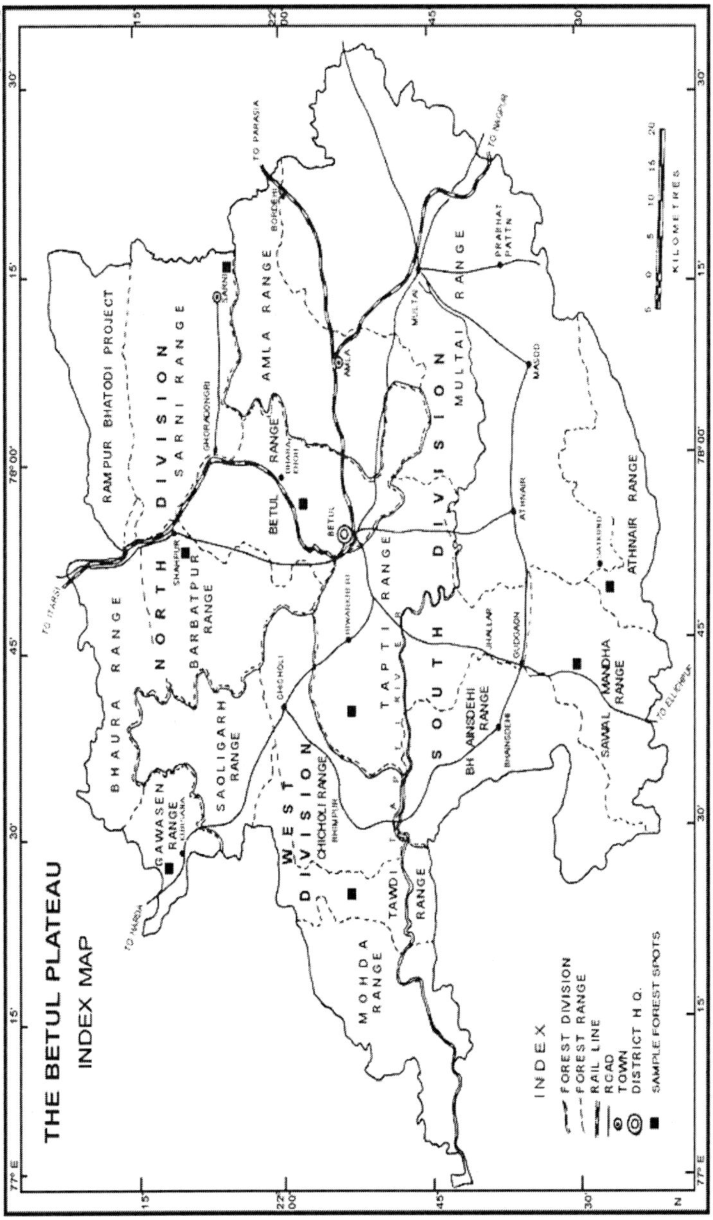
PLATE 1.2
THE BETUL PLATEAU
INDEX MAP
INDEX
FOREST DIVISION
FOREST RANGE
RAIL LINE
ROAD
TOWN
DISTRICT H.Q.
SAMPLE FOREST SPOTS
KILOMETRES
NORTH DIVISION
WEST DIVISION
SOUTH DIVISION
RAMPUR BHATODI PROJECT
BHAURA RANGE
SARNI RANGE
AMLA RANGE
BETUL RANGE
BARBATPUR RANGE
SAOLIGARH RANGE
GAWASEN RANGE
CHICHOLI RANGE
TAWDI RANGE
MOHDA RANGE
TAPTI RANGE
MULTAI RANGE
ATHNAIR RANGE
BHAINSDEHI RANGE
MANDHA RANGE
SAWAL
TO NAGPUR
TO PARASIA
TO ITARSI
TO ELLICHPUR

Geometric progression technique is adopted to prepare the distributional maps.

Transect Method

Hilly terrain where habitat features show steep gradients have been surveyed by the transect method. For the case studies of the medicinal plants in such environmental conditions, the hypsographic curvature of land has been considered in the selection of representative of ecological units. This method is usually employed by ecologists and is based on statistical analysis and gives qualitative information i.e. floristic enumeration, stratification, aspection, vegetative life forms etc.

In the study area, the vegetation sampling is made by studying a line across the area and recording the species that occur along the line. The data so recorded was then analysed. A purposive random sampling method proposed by

Stimson (1964, 164-168) and Murravy (1963, 155-162) has been adopted in the selection of spots for transect and quadrat method. Thus six sample cases in different ecological forests of this region have been selected and studied, of which only three have been described in the respective chapters.

Quadrat Method

For quantitative estimation of individuals of different species in valley area and plains the quadrat method proposed by Cotton and Curtis (1956, 451-460) has been employed with certain modification as per need of the terrain. In this method quadrat laid at random and every individual is faithfully recorded. For illustration purpose in the field, the quadrat of 10 metres x 10 metres is selected. Further the size of the next plot is doubled at a constant rate and number of species encountered in that quadrat are counted and noted until a point is reached where further increase in the size of sample plot did not add any additional information or number of species for determining the minimal size of a quadrat to study vegetation between the area of quadrat and the number of species occuring within it, a

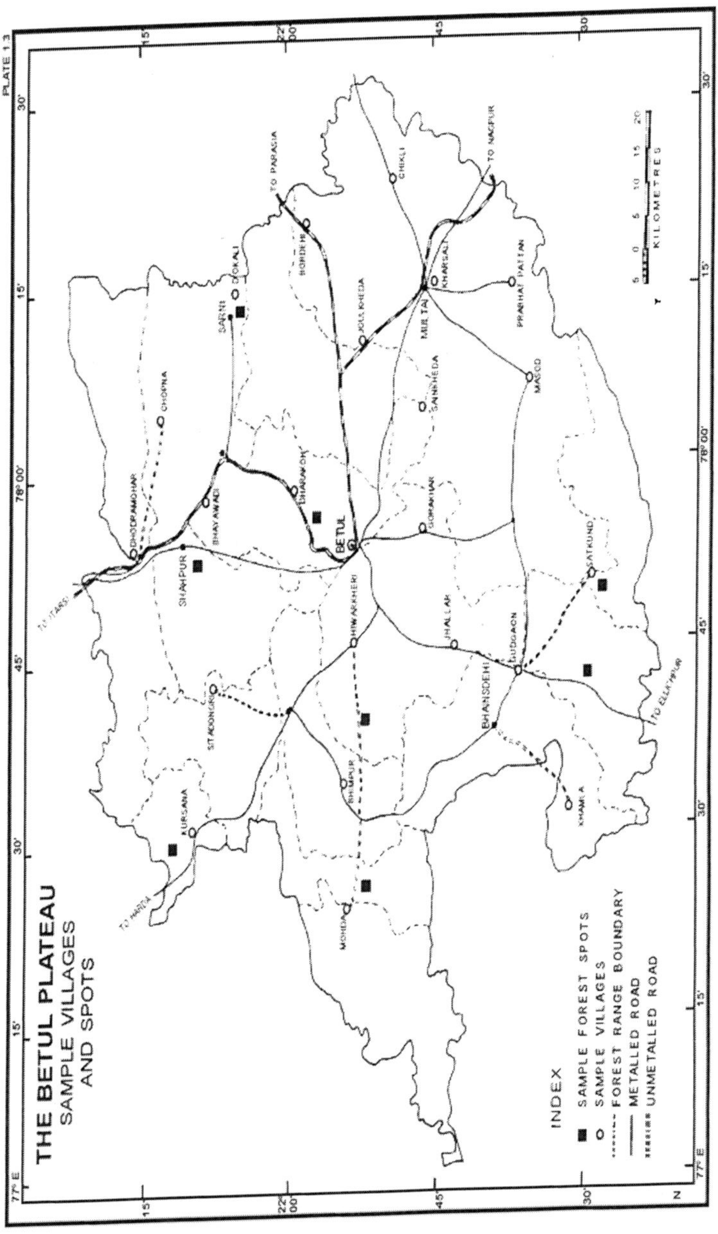
PLATE 1.3
THE BETUL PLATEAU
SAMPLE VILLAGES
AND SPOTS
INDEX
SAMPLE FOREST SPOTS
SAMPLE VILLAGES
FOREST RANGE BOUNDARY
METALLED ROAD
UNMETALLED ROAD
KILOMETRES
BETUL
MULTAI
TO NAGPUR
TO PARASIA
TO ITARSI
TO HARDA
TO ELLICHPUR
SHAHPUR
CHOPNA
SARNI
KHARSALI
PRABHAT PATTAN
MASOD
GORAKHAR
JHALLAR
GUDGAON
BHAINSDEHI
SATKUND
KHAMLA
BHIMPUR
AURSANA
CHIKLI
BORDEHI
BHAYAWADI
SITADONGRI
CHICHOLI

Density

It is the number of an individual species in one quadrat. The formula is as under:

$$\text{Density} = \frac{\text{Number of the individual plant species}}{\text{Total number of quadrats sampled}}$$

Relative Density

Density of one plant species expressed as the percentage of the densities of all plant species is called the relative density. It can be presented in the form of following formula:

$$\text{Relative Density} = \frac{\text{Density of an individual plant species}}{\text{Sum of density of all the plant species}} \times 100$$

Field Observation

Field observation has been done in order to get information about the crop condition, crop quality and plant community and their interrelationship with physical environment of that area.

Personal Interviews

Survey gives a review of interaction between man and his surrounding plant wealth. With a view to obtain first hand information about the use of plants, a field survey of twenty two villages of the study area is undertaken. The villages were arranged alphabatically. On the basis of their number 1.6 per cent villages situated in each of the tehsil are selected. For this purpose the systematic stratified random sampling method has been adopted.

This survey is conducted with a view to collect information about plants use and their application by the local and tribal people, as well as to investigate the distribuation of these species. Survey is based on the questionnaire method. In order to obtain various information concerning the plants used for different purpose, a proforma to incorporate all the information including their geographical distribution, potentials and problems were prepared. Besides some relevent information are also collected from foresters, inhabitants and herbalists through personal interview. For this purpose 84 persons are interviewed. Further in this context similarly the observations revealed that these useful plants can obviously be divided under five main categories which are known as applied categories. Thus, applied catergorization of the useful plant species has been done according to the nature of thier utilization for the welfare of human beings as well as domestic animals namely, (i) Commercial 24 per cent (ii) Medicinal 37 per cent (iii) Fuel 21 per cent (iv) Fooder 12 per cent and (v) Edible 6 per cent.

The species have been grouped into two classes, in order to fatch ease in discussion, (i) those informed by 0.5 to 20 per cent of the respondents (ii) those informed by more than 15 per cent of the respondents. The species of the latter class has been comprehensively discussed as compared to those classed under the former class. This arrangement has been preferred keeping in view the usefulness of the species.

REFERENCES

Aikman, L., 1974, Nature's Gift to Medicine *National Geography,* vol. 146 (3), 420-400.

Agarker, S.P., 1929, Vegetation and External Factors, *Journal of the Maedras Geographical Association,* vol. 6 (2), 29-35.

Ayensci, S. Edward, 1979, *Plants of Medicinal uses with special reference to Arid zone Thar desert and land Plant Resources,* Texas Tech. University, Texas, 117-148.

Barry, C., 1977, *Biogeography-An Ecological and Evolutionary Approach*, Oxford, Cox. Blockwell.

Bharucha, F.R., and Meher-Homji, V.M., 1965, On the Flora Elements of the semi-Arid zones of India and their ecological significance, *The New Phytologist*, 64, 330-342.

Black, C.A., 1957, *Soil, Plant Relationships*, New York, John Willey & Sons.

Bose, S.C., 1948, Plant Geography of Damoder Valley, *Geographycal Review of India,* Vol. 1 (9), 112-137.

Cain, S.A., 1971, *Foundations of Plant Geography*, New York, Harpar and Bros.

Champion, H.G., 1936, A Preliminary Study of forests Types of India and Burma, *Indian Forest Research Journal*, Vol. 1 (i), 1-286.

Champion, H.G., and Seth S.K., 1968, *A Revised Survey of the Forest Type of India*, New Delhi, Manager of Publications.

Charan, A.K., 1992, *Plant Geography,* Jaipur, Rawat Publications 225.

Chatterjee, D., 1939, Studies on the endemic flora of India and Burma, *Jour. Roy. As. Soc. Bengal Sci.* 5:19067.

Choubey, K., 1994, Allopathic and Traditional Health Care Services to Tribal Population of Bastar (M.P.), *Geographical Review of India Jour.* The Geographical Society of India, Calcutta, Vol. 56, June (2), 73.

Clark, C.B., 1898, *Subareas of British Empire,* illustrated by the detailed distribution of Cyperaceac in the Empire, Jour. Linn. Soc. London, 34, 1-14.

Cottam, G. and Curtis, J.T., 1956, The Use of Distance Measure in Phytosociological Sampling, *Ecology Jour.* Vol. 37, 451-460.

Croizar, L., 1952, *Manual of Phytogeography,* D.W. Junk, The Hague, 587.

Dansereau, P., 1957, *Biogeography, An Ecological Perspective,* New York, Ronald Press.

Dansereau, P. and Arreos, J.T., 1959, Essais d' application de la dimension structural en phytosociologie, *Vegetation.* 9, 48-99.

Darwin, C., 1859,Origin of Species by means of Natural Selection, or the Preservation of Favoured Races in the Struggle for Life, London.

Das, R.B. and Sarup, S., 1951, The Biological Spectrum of the Indian Desert Flora. *University Rajasthan Studies,* Jodhpur, 1, 36-42.

Dassman, R.F., 1968, *Envioronment Conservation,* John Willey and Sons, Inc, New York, Second ed; 173.

Davis, Petter, 1979, *Science in Geography,* Vol. 3: Data Description and Presentation, Oxford University Press, London : 60-61.

Deolia, R.K., 1994, *Medicinal Floral Ecology in Central India,* Classical Publishing Company, New Delhi.

Dev, D.B., 1960, Forest Types: Study in Manipur, *The Indian forester,* Vol, 86 (2): 94-111.

Drude, O., 1890, *Handhuch der Pflanzengeographic, Stuttgart Engelhorm:* 582.

Gaussen, H., 1959, The Vegetation Maps, Travaux de la section Scientifique Technique Tome I, Fascicule IV, Pondichery, Institute Francais.

Good, Ronald, 1947, *The Geography of the Flowering Plants,* London, Liongmand Green and Copp. 18-20.

Gurg, R.K. and Bandhu, D., 1986, Tribals and Forests, in Rajasthan, in Bandhu and Gurg (ed.) *Social Forestry and Tribal Development,* I E S Publication, 151-162.

Haggett, P. , Cliff, A.D., and Allen Frey, 1997, *Locational Models,* London, Edward Arnold.

Hanson, H.C. and Churchill, E. D., 1965, *The Plant Geography,* London, Oxford Press, 16.

Hartshorne, Richard, 1939, *The Nature of Geography,* Pennsylvania, the Association Lancaster : 35-130.

Hooker, J.D. at et al., 1879, *The Flora of British India,* I-vii, London.

Hooker, J.D. and Thompson, T., 1855, *Flora India,* London.

Hooker, J.D., 1906, *A Sketch of the flora of British India,* London, Oxford Press.

Humboldt, A. Von 1807, Ideen Zaeiner Geographic der Pflanzen nebest einem Naturgem aldeder Tropenlander, Tubingen.

Johnston, M.C., 1979, Medicinal Plants of Southern United States, *Arid Land Plant Resources,* Texas Tech. University, 179-185.

Kambush, A., 1924, The Forest of South *India Journal of the Madras Geographical association,* Vol. 2 (i), 15-35.

Karnik, C.R., 1956, Flora of Tapti Valley, *Bombay Geographical Magazine.*

Khan, A.H., 1935, Forest of Anantapur, District, *Journal of the Madras Geographical Association,* Vol. 9 (a), 39-48.

Kunkle, S.H. and Thames, J.L., 1977, Guidelines for Watershed Management, F.A.O., Rome.

Luthara, J.C., 1953, Community Grouping Plant under Geographical Variation, *Bulletin of the National Geographical Society of India,* Vol. 18, 11-21.

Maheshwari, J.K., 1980, Eco-Development of Narmade Valley in Man, *Development and Environment,* ed. by R.S. Doria, New Delhi, A.P.H.

Mani, M.S., 1974, *Ecology and Biography in India,* Dr. W. Junk. b. v. Publishers, The Hangue.

Margalef, Dr., 1958, *Perspective in Ecological Theory,* Chicago, University press.

Miiller, P., 1976, Biogeography : As a means Evaluating Living Spaces, *Applied Sciences and Development* (7), &-18.

Murray, Chapmen, 1963, Sample Method for Geographers, *Pacific Viewpoint,* Vol. 4 (2), 155-166.

NRSA., 1980, National Remote Sensing Agency by Department of Space Government of India.

Oommachan, M. and Masih, S.K., 1991, Conservatiuon and Cultivation of Excessively Exploited and vanishing Medicinal Plants of Madhya Pradesh, Agarwal, G.P. and Hasija, S.K. (Ed.) *Focal Theme (Botany)* ISCA Symposium, Delhi, Narendra Publishing House: 197-212.

Rao, A.S., 1968, The Vegetation of Khasi and Jaintia Proceedings of the Pre-Congress Symposium and Meghalaya and Eastern Himalayas, held at Gauhati, International Geographical Union. Department of Geography, Gauhati University.

Raunkiaer, C., 1907, The Life- Form of Plants and their bearing on Geography, (in his collected essays published) Oxford, Clarendon Press: 2 - 104.

Raunkiaer, C., 1934, *The Life Forms of the Plants and Statistical Plants Geography,* Oxford, Clarendon Press.

Razi, B.A., 1955, The Phytogeography of the Mysore Hill Tops, *Jour. Mysore Univ.,* 14, 87-107.

Puri, G.S., 1980, *Indian Forest Ecology,* 2 Vols, London, Oxford Press.

Pyle, G.E., 1979, *Applied Medical Geography,* London, Oxford Publication.

Schimper, A.S.W., 1903, *Plant Geography,* Oxford.

Shaileshwari, E.L., 1916, The Oak-Hom Bean Woods of Hart fordshair, *Ecology Journal,* Vol. 4, 83-117.

Sharma, D.G., 1967, Te Problem of forest Protection in India with special reference to Madhya Pradesh, *The Indian Forester,* Vol. 93 (6), 407-410.

Sharma, S.K., 1971, Forest Resource and its conservation in the Baghelkhand Plateau, M.P. Allahabad, *National Geographer,* Vol. 6, 65-73.

Singh, G.S., 1982, *Forest Ecology of India,* Jaipur, Rawat Publications.

Stimson Catherine, D.J., 1964, Sample studies their contribution to Geographical Education, *Australian Geographer,* Vol. 9 (3), 164-168.

Vishwnathan, G., 1931-32, Type of Forest Flora in Malaber, *Journal of The Madras Geographical Association.*

Varma, S.K., 1967, Vegetation types of Jaipur, *The Indian Forester,* Vol. 93 (i), 7-17.

Watts, D., 1971, *Principles of Biogeography,* London, Mc Graw Hill.

2

Environment and Vegetation

Environment is such a type of structural profile which relates organisms and their set of surroundings. Organisms in environment behave like a part of system. Environment is defined more comprehensively by other as holistic view of the world as it functions at any point of time with a multiplication spatial elemental and socio-economic system distinguished by quality and attributes of space and mode of behavior of biotic and abiotic forms. *(Dikshit, 1984, 66)* Environment is an inseparable whole and is constituted by the interacting system of physical, biological and cultural elements which are inter]inked individually as well as collectively in myriad ways. *(Singh & Dubey 1983, 73)*

Natural vegetation of region is a prime factor as a determinative of environment, connecting physical components with biotic components and is directly affected by each of them. The physical components i.e. land forms, rocks, water bodies, soil and climate determine the variable characters of the floral habitat, its opportunities as well as limitations. Moreover, in turn man as the most powerful biotic constituent also does modify continuously the spatial attributes of vegetation. Before undertaking the study of regional flora, the study of the mode of operation of environmental and biological process and the nature of mutual interaction between them must be studied. Keeping this fact in view this chapter is provided with physical and socio environmental profiles.

PHYSICAL ENVIRONMENT

Physical environment creates the floral profile variation. Since various environmenta1 factors affect the life of the plants in a holistic manner, it becomes difficult to understand this mechanism of the nature of influence of individual factor. To understand this mechanism of the environmental influence in the study area, it thus becomes essential to study the effect of each factor separately. Hence, a detailed study of them is mentioned in the following groups:

1. Geological Profile.
2. Physiographic Profile.
3. Drainage.
4. Climatic Profile.
5. Edaphic Profile.

GEOLOGY AND PLANTS

Geology has a profound impact on the nature *(Nagi, 1986, 01)*. It plays a significant role in the formation of soil and vegetation of a region. Soil provides a solid, yet penetrable foundation for plant growth. Thus, the plant growth is determined to a considerable extent by the amount of nutrients in the soil which depends on nature of underlying rocks and is another factor which influences depth of a root system, size and erectness of plant, vigor of the vegetative organs and woodiness of stem.

Approximately one fifth area is represented by the rocks of Archaean era. Remaining area is made of different rocks of later geological ages. The following table shows the age assigned to various formation of the study area.

In the study area, the distribution of rock formation shows that the oldest rock granite are exposed in the mid-eastern part of the study area, Gondwana rocks are found in the north-eastern part,

Table 2.1

Betul Plateau : Geological Time Scale

Geological Age	Formation
Recent	Soil
Pleistocene-Recent	Laterite
Upper Cretaceous-Eocene	Deccan Trap (Flows sills and dykes)
Cretaceous	Lameta beds
Jurassic	Upper Gondwana System: Jabalpur beds
Permian	Lower Gondwana System: Bijori, Motur, Barakar, Talchir
Archaean System	Basic intrusive, Schists and quartz, granite and gnesis.

while Deccan lava flows predominantly in the central, southern and western sections (Plate 2.1).

GEOLOGICAL FORMATION AND PLANT COMMUNITIE

Archaean System

The system falls in between Chicholi in the west and Tawa valley in the east. A broad tract of Archaean is occupied by the granite rocks. At the southern base of the Gawaligarh hills a very small exposure of metamorphic rocks occurs close to Salbardi. Here, the beds are only traced for 23 kilometres with a few hundreds metres in breadth. The granite in the Bel valley shows sign of considerable disturbances.

In the plateau, the Dharwar rocks which are folded with the gneisses are described to include phyllites, quartz, mica, schist's, quartzites and epidiorites.

Intrusive into the gneisses and granites of the plateau are large masses of basic rocks such as epidiorites, hornblends, schists, diabases and amphibolite in the form of sills and dykes. Theses were presumably, included as dykes and sills of dolerite and gabbro, remments of the origina1 phroxenic rocks, some of them with olivines

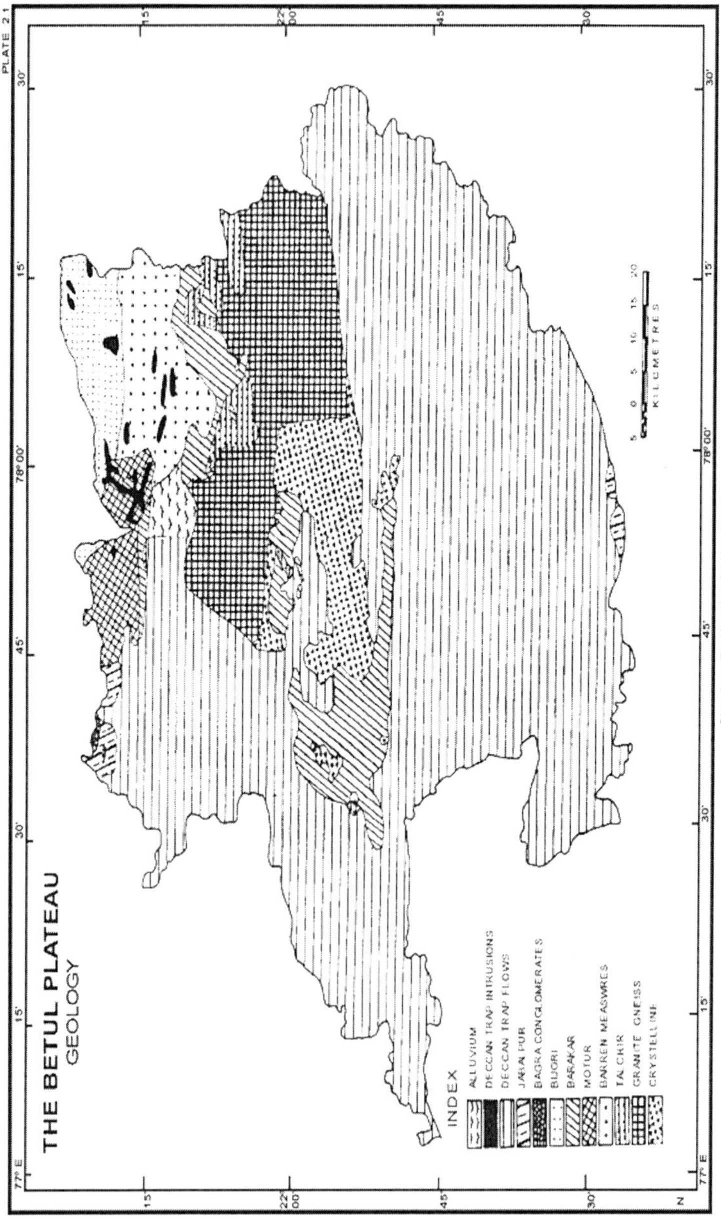

Source : Geological Map of India

being found here and there. The amphibolites are sometimes granitiferous especially when they have been subjected to shearing. The hornblends schists are often rich in a green pyroxene.

The main relationships between this system and distribution of vegetation can be underlined as follows:

Soil derived from metamorphic rocks are heavy and compact in structure with greater moisture retentivity, so these areas are mainly occupied by moist miscellaneous species of plant which have higher moisture requirements. In the study area, prominent forests of such species occur over this formation in the Betul and Amla forest ranges. The high Landi and Mowar plateau form the middle to eastern Betul plateau comprise; of metamorphic rocks and is covered with

moist mixed forests. The dry mixed forests are associated with metamorphic rocks spreading in southern boundary of the area.

Gondwana System

The rocks of the Gondwana System which comprise a succession of sandstone and shells with seams of coal lie over the metamorphic crystalline rocks of the Archaean system with a distinct and profound disordance. In these rocks, occur organic remains of fresh water origin. The beds are distributed in patches of various sizes, which are in a linear set and coincide more or less with the present river valley. The rocks in general dip 5 to 6 degrees to the north but often show sign of disturbances at place with steep dips on account of faults. The faults took place at the time of basalt flow and sills of the cretaceous time.

In the study area, the Gondwana System consist beds of Talchir, Barakar, Motur, Bijori, Bagra and Jabalpur.

The Talchir Bed

It is about 300 metres thick, consists of sandstone and green clays with boulders and is extremely rich in plant fossils. In the region of Sonada, thin coaly layers occur in these rocks with leaf impressions. On account of the homogeneous nature of Talchirs, the beding planes are often obscure. The rocks of this group are often bounded by faults. The age of these faults is uncertain but in some cases it is post-Gondwana for the Deccan trap flow which have been seen shifted by the same disturbances. The Talchir bed lies from the catchment areas of river Phopas in the east to Tawa valley.

To the west and north-west of Betul is a group of rocks which separate the Deccan traps from the Archaean rocks. The beds comprise a conglomerate sometimes cherty, with occasional sand stones and clay varying in thickness, throughout a considerable distance they appear not to exceed 30 metres and are often thinner. The conglomerate was referred to the Lametas, while the soft, argillaceous sandstone with harder bands and occasional red shales or clays shows considerable resemblance to some of the rocks but sometimes dips at steep angles apparently as a result of a fault. The rocks are, however referred to the Talchir series.

The Barakar Bed

The Barakar stage consists of brown and grey flaggy felspathic sandstones and shales with bed of coal. It is the only group in which coal is found. The Barakars are characterized almost everywhere by poorly preserved plant fossils, chiefly impression of leaves or stems of vertebrata. In Sonada coal certain horizons of these coal measures. The Barakar beds bearing a circular curve to the north and north east of the study area following roughly the coarses of the Tawa and Bhaura streams. They are bounded to the south by the older Talchirs and to the north by the younger Moturs.

The Motur Beds

The Barakar are succeeded without any noticeable break by the beds of the Motur stage which may be upto 1800 metres in thickness. The Motur rocks consist of thick, soft or coarse earthy, grey and brown sandstones with occasional tentacles of clay and shales. The Motur sandstones resting over the northern boundary of the study area. They lie from the east, near the meeting point of the river Tawa and Bharanga to the village Kalapani in the west.

The Bijori Beds

The Bijoris which are 180-240 metres thick comprise shales which are occasionally carbonaceous micaceous, flages and sandstones. The beds roll in places but dip eventually northward. The Bijori beds occupy the north-east boundary of the study area. They are exposed in the valley of the river Bharanga and Sonbhadra.

The Bagra Bed

The Bagra facies is very irregular in composition. Although largely composed of conglomerates, which are often coarse, it includes frequent bands of calcarious sand and variegated clay. They are connected by a matrix which is usually of a deep red colour. They are found in the Morand valley as a narrow belt toward the north-west portion of the area under study. The Bagra beds are bounded to the south by the Jabalpur beds.

The Jabalpur Beds

The Jabalpur beds consist of massive sandstones alternating with soft white clay. In addition, subordinate beds of conglomerate of earthy heamatite, carbonaceous shales, red clay and chert may occur. The thickness of these beds is between 75 to 150 metres, in the study area they are seen developed to the north west of the Duni river valley near Deoban and to the east of the Morand valley near Taramkhera by the researcher.

In the study area one interesting feature of the impact of geology on the distribution of vegetation is seen in Gondwana sandstone area. The sandstone consists of massive, soft sand stones, gravel beds, clays etc. The mixed forests are predominating over this formation. The underlying rocks of the Gondwana sandstones give birth to infertile soil which is occupied by dry miscellaneous forests.

Some of the Talchir and Barakar soil are more clayey and give a slightly better forest crop of moist miscellaneous species. Bijori's and Motur beds contain fertile soil. The former is occupied by teak and latter by Saj an associated species of mixed forests. In Jabalpur beds the sandstone country is capped by thin sandy soil which favours growth of dry mixed forests.

The Lameta Formation

The Lameta rocks are found to intervene between the trap flows and the underlying Gondwana rocks. The beds comprise of purple clays and are accumulation of volcanic agglomerate mixed with a little sand. The clay is composed of numerous angular quartz grains. They are exposed near Taramkhera and along the road east of Pat (Tableland). In lameta formation teak is invasive in the clear fellings of the mixed forests.

The Deccan Trap

The Deccan traps are the one major geological formation occurring in the area under study. More than half part of the plateau is covered by this formation. The extensive trap formations are found in the Betul-Multai plateau, Bhainsdehi and Morand plateau. By far the most extensive rocks in the study area are, the trap flows, dykes and sills. The flow is not uniform in a section. The base consists of a thin porous layer of earthing basalt. In typical cases, this part shows vertical columnar joining. Above the main mass is a thick layer of tough basalt characterized by its conchoidal fracture.

The Deccan trap is the name applied to great volcanic formations, resulting from a series of eruptions from fissure and cracks in the surface of the earth *(West 1959, 44-52)*. In the flows of these formations a fine grained irregularly jointed flow weathers into spherules medium grained black basalt, a porphyritic flow and a very mygdaloidal flow have been recognized in the Betul plateau.

On trap flows or intrusions teak predominates often in the virtual exclusion of other tree species as the soil has sufficient minerals to support a fairly good forest crop. The associated areas of trap are hilly and densely forested. The trappean areas with dry conditions are covered with Salai forests.

Laterite

The Deccan trap shows in places a thin cap of Laterite which is of the vescicular type. It usually forms porous hard reddish capping over the basaltic rocks. It is composed of the mixture of the hydrated oxides of alumina and iron with the other oxides like manganese and titanium. Small patches of Laterite soil are seen in western and southern most part of the study area. They favour growth of Sawana type vegetation.

Recent Deposits (Alluvium)

The northern part of the study area i.e. the middle Tawa valley plain is covered by recent alluvial deposits of yellowish brown clay with numerous intercalated bands of sand and gravel. The thickness of the alluvial deposits exposed along the banks of the rivers usually does not exceed 35 metres in depth. The valley of Machna is also covered with thick alluvial deposits.

The alluvial supports the growth of moist deciduous teak and mixed forests, so the Machna and Tawa valley are extensively covered by such forests.

In the study area, by field investigation it can be concluded that geology has played a significant role in the distribution of forests. A particular forests type has been noticed over a particular rock formation. Soil derived from deccan trap favours teak often to the almost exclusion of all other species so that it is intimately associated with and thrives exclusively on the Deccan trap towards the western and southern parts of the study area. Soil derived from sandstone and metamorphic rock formations are invariably under the mixed type of forests.

TOPOGRAPHY AND PLANT ECOLOGY

Topographic features are deemed to be the cradle for the evolution of plants. The topography shapes the forests crop. Thus, the study of general relief and the total landscape are of paramount importance in an ecological study as they determine the distribution of plant communities *(Sharma 1988, 92)*. In view of the above statement the detailed study of topography has become a prerequisite which is given below.

The study area is a high land surrounded by high Satpura range. Its general elevation is about 609 metres. It is bounded by the Mahadev range to the north, the Kalibheet range to the north-west and the Gawaligath hills to the south. The range falls steeply towards the south in the Barar plain and to the north in the Narmda valley. Towards the east, the area is separated from the Chhindwara plateau by a hilly sloped Kanhan valley. The plateau is high in the east, gradually lowering toward the west and makes a slope from south to the north and west (Plate 2.2).

The topography of the plateau is dissected one. The drainage seems to be the most dominant factor of the plateau; hence a detailed and critical study of drainage is done separately.

The topography of the area under study can be segmented as follows (Plate 2.3) :

(i) THE SATPURA RANGE

1. The Mahadev Hills
2. The Kalibheet Hills
3. The Gawaligarh Hills

(ii) THE CENTRAL PLATEAU

1. The Betul - Multai Plateau
2. The Chicholi - Morand Plateau
3. The Bhainsdehi - Kamla Plateau

(iii) THE VALLEYS

1. The Tawa River Valley
2. The Tapti River Valley

THE SATPURA RANGE

The Satpura range is a predominating feature of the physiography and geography of the region. It surrounds the study area on three sides. The trends of ancient folds are in a North West

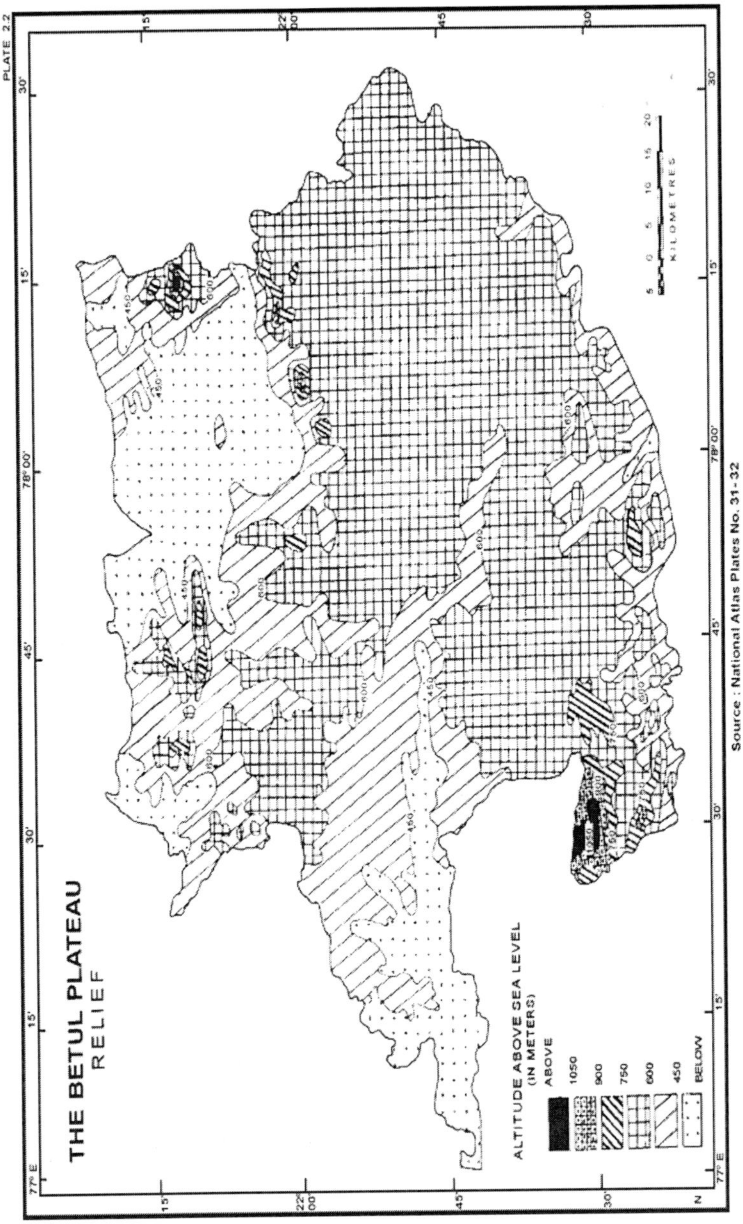
PLATE 2.2
THE BETUL PLATEAU
RELIEF
ALTITUDE ABOVE SEA LEVEL
(IN METERS)
ABOVE
1050
900
750
600
450
BELOW
KILOMETRES
5 0 5 10 15 20
77° E
15'
30'
45'
78° 00'
22° 00'
N
Source : National Atlas Plates No. 31-32

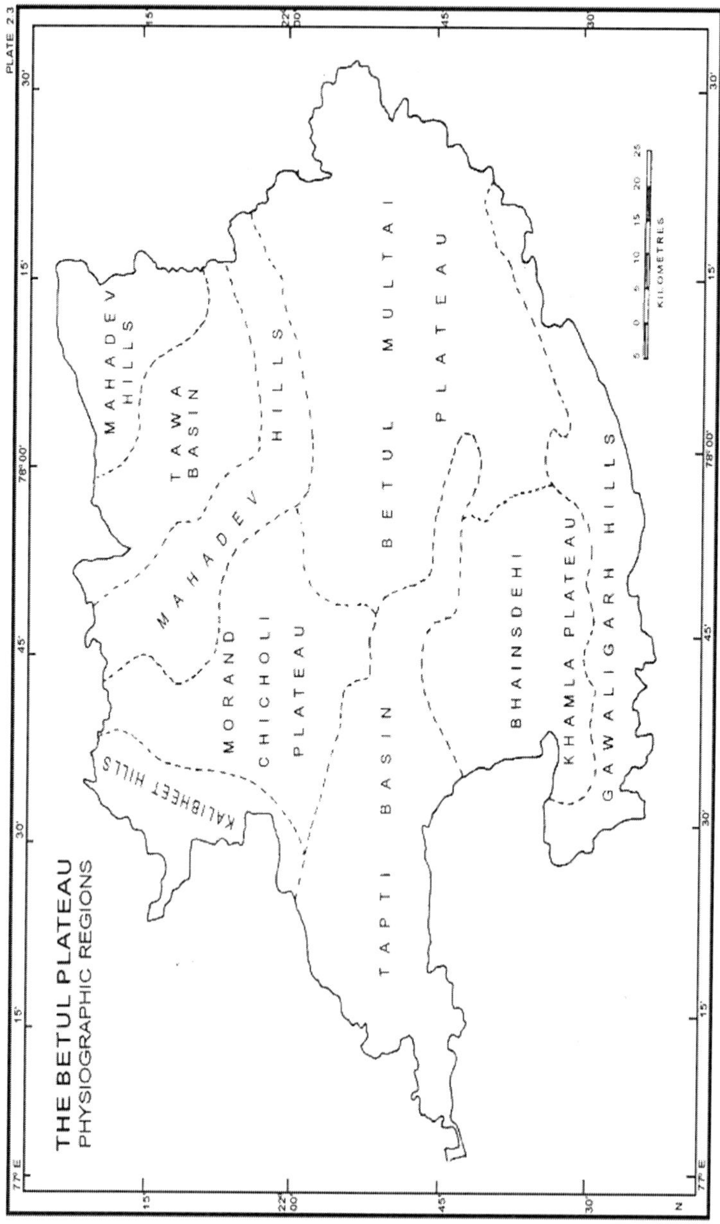
PLATE 2.3
THE BETUL PLATEAU
PHYSIOGRAPHIC REGIONS
MAHADEV HILLS
TAWA BASIN
MAHADEV HILLS
BETUL MULTAI PLATEAU
MORAND CHICHOLI PLATEAU
KALIBHET HILLS
TAPTI BASIN
BHAINSDEHI
KHAMLA PLATEAU
GAWALIGARH HILLS
KILOMETRES
Source : National Atlas Plates No. 31-32

and South direction. The highest surface develops partly on the upper Gondwana sandstone and partly of trap rocks. The northern part of the range develops on upper Gondwana sandstone but southern and western parts of range are extended on trappean hills. The range is known by different names in part of the plateau which are described as follows:

1. **The Mahadev Hills** : The western extension of the Mahadev hills lies to the northern part of the plateau and is stretched from north-east to Morand valley to the west.

   These hills overlooking the Narmada trough northwards but cut off to the south by a stupendous line of cliffs, over looking the low hills *(Singh, 1971, 569)*. The hills attend an average height of 700 metres. They are composed of sandstone and partly of crystalline rocks. The upper Gondwana sandstone of the Mahadev series on weathering take peculiar forms resembling bastions, buttresses and battlements. The main hills of the range are cut by the Tawa and other tributaries of Narmada system and broken up into separate hills which are as follows:

   (i) ***The Asir Hills*** : These hills are extended to the north-east boundary of the study area and form the northern spur of the main range. The spur scarps in the Tawa valley toward the south about 152 metres. The hills have an average height of 600 metres and are highly dissected with number of ugly looking conceous gullies.

   (ii) ***The Keelandeo Hills*** : These hills form the eastern boundary between the Tawa and Bharanga River. They have scarping edges and stand above the Tawa valley. The Keelandeo hills rise to over 640 metres from the valley of which Keelandeo forms the highest (1107 metres) peak in the northern and central parts of the study area.

   (iii) ***The Bhanwargarh Hills*** : The Bhanwargarh hills form scarpment of the study area towards the Narmada valley and are extending from the west of the Tawa

River to the Morand plateau. The hills rise to about 420 metres from the Tawa valley and attend the maximum height of 700 metres in general.

(iv) ***The Southern Range*** : The range lies to the south of Tawa River and runs in the east-west direction. It has steep hills and scarping in the Tawa valley. The important peaks of it are Bharkach, Matyardeo (911 metres), Paraskot (919 metres) and Sirrikot (828 metres).

The eroded land of Asir hill has dry type mixed forests whereas its peaks and scarpments have been occupied by Savana type vegetation. The lower slopes of this hill have moist type forests and bamboo forests. The Keelandeo hills have been covered with subtropical wet forests because of its higher elevation and micro climatic conditions. The Bhanwargarh hills are heavily clothed with teak forests. The peaks and ridges of this hill favour the growth of Salai forests whereas its northern scarpment is covered with mixed forests. The southern hills of Mahadev range consist of moist mixed forests while lower slopes show occurrence of bamboo forests.

2. **The Kalibheet Hills :** These hills are extended toward the north-west of the plateau as horseshoe shape. The Kalibheet trappean hills to the north have been dissected into isolated or groups of masses. It is eroded by the Morand River and its tributaries which rises upto 715 metres. Towards the northern part of these hills, the Saoligarh hills (679 metres) and Narwargarh hills (784 metres) are located and towards the south the Tapasari hills (821 metres) and the Alampur hills (758 metres) are extended. It has a chain of main peaks in the northern part of the Tapasari, the Chouradev and Phoplia (784 metres).

The hills, scarpment and peaks of this hill are covered with forests. The shallow sandy soil areas consist of dry teak forests but in crystalline areas especially on hill tops the conditions are dry and so dry teak forests are replaced by

Salai forests. On slopes, the patches of Bamboo forests can be seen.

3. **The Gawaligarh Hills** : The series of hills lies to the Southern most boundary of the plateau from west to east in the form of range. These hills rise to 975 metres and present a steep slope to the south, part of which is definitely a fault scarp. Its general elevation declining towards the east and lost in the plateau. The peak (1137 metres) in the west of Khamla is the highest point in the whole of the study area and forms part of the Gawaligarh hills. Dhar (1114 mtrs.) and other peaks in the east of Khamla also attaining the same height. Other important peaks of this range are Athner (915 metres) and Satkund (803 metres). The southern range of the hils over looking the Barar plain. The highest peaks of the southern range are Salbardi (941 metres) and Kalapakhan (777 metres).

   The range is popularly known for ecological class of floristic composition. The western higher elevations of these hills have been occupied by wet forests and eastern elevation is characterized by dry teak forests. Dry mixed forests occur on metamorphic formation and are confined to the southern slopes toward the east. The southern scarp is not much conducive to the growth of dense plant cover, therefore this part consists of scattered Salai forests.

CENTRAL PLATEAU

This plateau is surrounded by the Satpura range on three sides and forms a compact shape almost a square. The plateau is divided towards the north by the valleys of tributaries of Narmada and towards the south by the valleys of Tapti and its tributaries. Thus, eroded by rivers, the plateau is divided in to following sub-divisions:

1. **The Betul-Multai Plateau :** This plateau is high toward, the east. The highest range of the Satpura either scarps the northern rim of this high plateau. It spreads in a wide area from the Tawa valley to Betul Bazar in the west, to the

Kanhan valley in the east and from the Pattan in the south to the foot of Sirrikot peak in the north. It has a general elevation of 750 metres.

The axis of the plateau is a trap covered range over 762 metres high from the mean sea level and running almost from east to west. Its northern slope is drained by the Bel which flows to the east. The southern slope is drained by the Tapti, a sacred river. Its source is on the slope of the 791 metres high hill which is located to the north of Multai. This hill marks the water partings amidst the water of the Wardha and the Tapti to the south-east and south-west respectively and Kharpura Nala to the north. This part of the plateau is almost mentioned as the Multai plateau. This high mass of the plateau subsides into the fringing ravenous country of the Wardha and the tributaries of the Kanhan towards the north-west and east. It is probably this phenomenon which demarcates the boundaries of two plateaus, Betul and Chhindwara in the part. To the west, it lowers gradually and spreads in a partly undulating plateau region. The northern extension of the plateau is marked by the low range of hills.

The plateau is a wide-agricultural area due to black cotton soil but the northern and southern bordering areas of it are forested. The northern high plateau area adjacent to the Satpura range is covered with moist mixed forests and moist teak forests while the southern boundary consists of dry deciduous teak forests.

2. **Chicholi-Morand Plateau :** This triangular plateau lies to the western part of the study area of which southern part is broad while northern part is narrower. It spreads from the western extension of the Mahadev hills in the north to the Tapti valley in the south and from the Kalibheet hills in the west to the Bhanwargarh hills in the east.

   This high land marks water partings of the tributaries of Narmada, Morand and Bhangi in the north and the tributaries of Tapti in the south. It has a general elevation of about 640 metres. Its northern part is drained by Morand, Bhangi whereas southern part is drained by the Labada, Nishan and Betul.

The part of the plateau near Chicholi settlement is widely used for cultivation. The south-eastern plateau area consists of dry mixed forests while the western part with unfertile soil is covered with dry teak forests. Mixed forests and Bamboo forests are found in patches among the extensive teak forests along the valleys in the northern Morand plateau.

3. **Bhainsdehi-Khamala Plateau :** The south-western part of the study area comes under Bhainsdehi-Khamala plateau. Its northern boundary is marked by the Tapti and southern boundary is formed by the Gawaligarh hills. It forms a narrow belt bordered by the ravines of the Wardha and its Tributaries (Maru) to the east which separates the plateau from Multai plateau. The general elevation of the plateau is about 640 metres. In the eastern part it is about 701 metres in the west with large area rising above 762 metres. To the west around the Bhainsdehi it is much wider than around the Masod. The undulating part of the plateau drains into the Tapti. The southern alienated water divide is dissected on the southern slope although the Puma also taps the water of Bhainsdehi area and flows towards the south cutting across the highest part of the plateau. To the south of Athnair the plateau has been much eroded and lowered by up streams of the main river.

   The southern slopes of the plateau are almost wholly covered with forests. Dry teak forests can be seen over the Neelgarh hills and along the Tapti valley. The high Khamla plateau to the south is covered with dry mixed forests. This upland is a good place for floral life (herbal).

THE VALLEYS

1. **The Tawa Valley :** The Tawa valley lies at an elevation of about 396 metres above mean sea level between the peaks

of Keelandeo and Bhanwargarh. It is bordered by the cliff over a 304 metres high on the north east, south, west and south-east. The general slope of the valley is toward the north-west. The country is undulating with presence of a few residual hillocks and intersected by a large number of streams joining the Tawa. The soil mental is thin, poor and sandy except along the large water course where it is deep and of finer texture.

A large portion of valley is covered with forests favouring the growth of valuable trees to perfection on sites with better soil. The northern parts of the valley and the best soil elsewhere have been occupied for agriculture. Belts of moist deciduous mixed forests are found in the south eastern part while dry deciduous mixed forests can be seen toward the north-east part of the valley. The western portion of the valley is covered with teak forests.

2. **The Tapti Valley** : This valley lies to the western part of the study area and forming a triangular structure. It spreads to the east-west and is bordered by the Chicholi and the Bhainsdehi plateau to the north and south respectively. The 450 metres high contour distinguishes the Tapti valley from the plateau. Towards the east it has a narrow plain land bounded by the Neelgarh hills which becomes wider towards the west. Its general slope is from east to west. On the bank of the river with patches of alluvial soil, the land is limited in settlement. The valleys of the Tapti and its tributaries are bordered by the ravines which are covered with forests. The alluvial patches of the valley consist of dense mixed forests.

The physiography of the study area is characterised by moderate to steep slopes, deep valley, hills and undulating plateau areas. Soil, on scarping slopes is base deficient, hence these areas are very often supporting only stunted and scattered miscellaneous species. On the other hand deep slopes are relatively gentle, giving rise to deep soil and bear good vegetation in association with miscellaneous species. The higher elevation consists of wet forests whereas lower slopes are covered with moist type of forests. The hill tops and exposed slopes are occupied by poor quality type of forests.

DRAINAGE AND VEGETATION

The characteristics of the drainage particularly the frequency and density of streams are the determinants of the nature of terrain as well as the run off, depth of water table and aquifer. They also determine the depth and characteristics of soils and availability of moisture in it. They are, therefore, one of the major controlling factors of growth, distribution and composition of natural vegetation.

The drainage system of the study area can be grouped as follows:

1. The Arabian Sea Drainage System.
2. The Bay of Bengal Drainage System.

THE ARABIAN SEA SYSTEM

The drainage system includes the Narmada system and the Tapti system. The Tawa and the Morand are the tributaries of the Narmada system which flow towards the north. The northern and the central part of the plateau are drained into Narmada. The Tapti system drains the western and the south central part of the Betul plateau (Plate 2.4). The system drains near about 7782 square kilometres which is about 77.7 per cent of the total area of the plateau (Table 2.2).

The Tawa River

The river Tawa is one of the tributaries of the Narmada flowing in the north-eastern part of the study area. It rises in Chhindwara plateau. After making a course of about 32 kms. between two ranges of the Satpura hills, it enters the plateau on its confluence with the Baradha nala. It flows generally towards the west direction for about 35 kilometres in the plateau and receives the water of the Phopas

Table 2.2

Betul Plateau: Particulars of Tributaries

Name of River System	Name of Tributaries	Catchments area (Sq.Kms.)	Percentage of Catchment area (%)	Length (Kms.)
Narmada	Tawa	1844	18.3	60
	Sampna	145	1.4	18
	Machna	713	7.1	94
	Morand	1003	10.0	80
Tapti	Tapti	3702	37.1	181
	Purna	375	3.8	42
Godavari	Bel	784	7.6	43
	Maru	750	7.3	35
	Wardha	752	7.4	31
Total		10068	100	

Source: Jila Sankhyiki Pustika, 1991.

and the Daryakho nalas from the south, a few kilometres to the north of Ghodadongri. The river takes a northenly course from here and joins the Narmada. The total length of the river is about 185 kilometres out of which 60 kilometres are shared by the study region.

The Tawa basin is significant from the floristic point of view. The alluvial deposition along the river bed consists of good forest crop which includes all the main classes of the plant kingdom.

The Morand-River

The Morand is another tributary of Narmada flowing in the northwestern part of the study area. The catchment area of the river

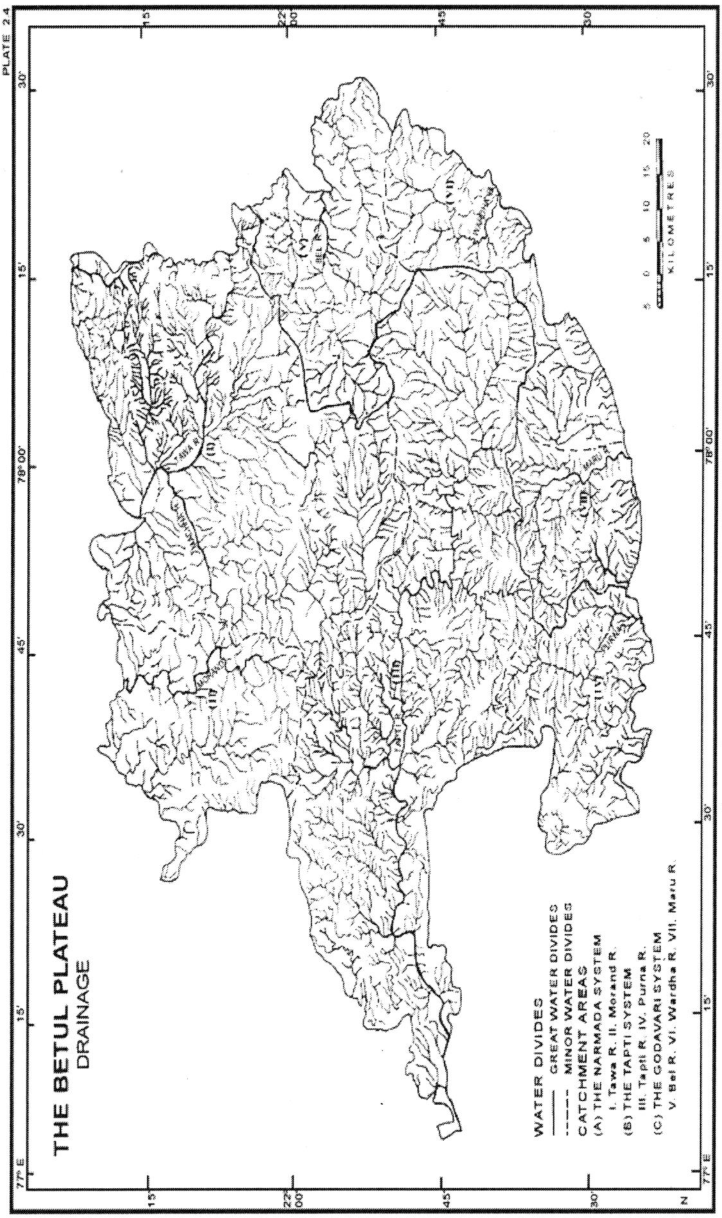
PLATE 2.4
THE BETUL PLATEAU
DRAINAGE
WATER DIVIDES
GREAT WATER DIVIDES
MINOR WATER DIVIDES
CATCHMENT AREAS
(A) THE NARMADA SYSTEM
I. Tawa R. II. Morand R.
(B) THE TAPTI SYSTEM
III. Tapti R. IV. Purna R.
(C) THE GODAVARI SYSTEM
V. Bel R. VI. Wardha R. VII. Maru R.
KILOMETRES
Source : Quarter inch Toposheets No. 55F, 55G, 55J, 55K

is 1003 square kilometres. It runs about 80 kilometres towards the north-on the Morand plateau.

The physical conditions of narrow valley of the Morand River are suitable for the growth of a variety of floral life. Along the river, the teak grows with other associated species whereas lower slopes such as Tanda nala have been occupied by Bamboo forests.

The Tapti River

It is one of the biggest rivers of western India. Its source is on 790 metres high hill and is located to the north of Multai tehsil. It takes a course of 25 kilometres towards the south-west and 50 kilometres towards the northwest making a semicircular curve. From here, it flows toward the western part of the study area, then plunging into a rocky gorge of Kalibheet and Chikalda ranges. After a course of about 181 kilometres from its source, it enters the Nimar District. The important tributaries of the Tapti in the study area are the Ambhora, the Khandu and the Kursi on the left bank and receive water from the Bhainsdehi and Multai plateau. On the right bank, its tributaries the Betul, the Batki, the Labadi and the Nishan drain the southern Chicholi plateau. The catchrnent area of the Tapti is near about 3702 square kilometres. In terms of percentage the area is about 37.1 per cent of the total area of the plateau.

The alluvial deposition of the Tapti valley is an important factor in governing the growth and distribution of vegetation. The upper part of the valley consists of Bamboo forests in patches while the middle part has mixed forests. Teak forests are also found in lower parts of the valley.

THE BAY OF BENGAL SYSTEM

In the study area, the Bay of Bengal drainage system consists of the Godavari system to the east. The main river of the system is Wardha it raises from the southern slope of Dahawadhana peak (811 metres) located at the north east of Multai. It flows for about 31 kilometres in the high plateau and drains a small area in the south-east. After a course it meets the Wanganga River which later joins the Godawari. The river Maru, flows towards the south and join the Wardha by marking a course of about 35 kilometres in the study area.

Due to the intensive use of land for cultivation quite scanty trees and other vegetation are noticed in the area of this drainage system.

Drainage Density and Frequency

The drainage density and frequency provide an expression of those elements of the terrain of an area which determine its hospitability to the growth of vegetation. A high density means heavy dissection of the terrain making it unsuitable for cultivation, but suitable for the growth of wild plants and vegetation. High drainage density may also suggest greater nearness of the interfluvial tracts to the drainage lines and thus greater accessibility of plants to the source of moisture, viz streams and nala. Contrarily low drainage density indicates smaller dissection of land which would make it more suitable for cultivation and would provide lesser accessibility of interfluvial areas to the local sources of moistures available in the adjoining streams. As such spatial variation in the drainage density and frequency of streams may suggest the degree of hospitability of those areas to the variations in vegetative growth.

In the study area, density varies from 1 to 28 kilometres for each 5 square kilometres of area (Plate 2.5). It is clear in the plate that the density increases toward east-west and north-east. In the region, the Asir range, settlement near Betul, the southern part of Athnair, northern Bhainsdehi plateau and southern Chicholi plateau are the areas with very high density (28 and above). The major portion of the study area falls under high density (21-28) region including the Tapti Basin, the Bhainsdehi plateau and the Tawa valley. Very low density (1-7) is found in narrow belt of east, south eastern and southern bordering areas. The rest of the area has low (7-14) and medium (14-21) density. Low density is marked in the northern Morand plateau, the southern Tawa basin and the Landi plateau while medium density areas are the Central Morand plateau the Betul-Multai plateau and the high Khamla plateau.

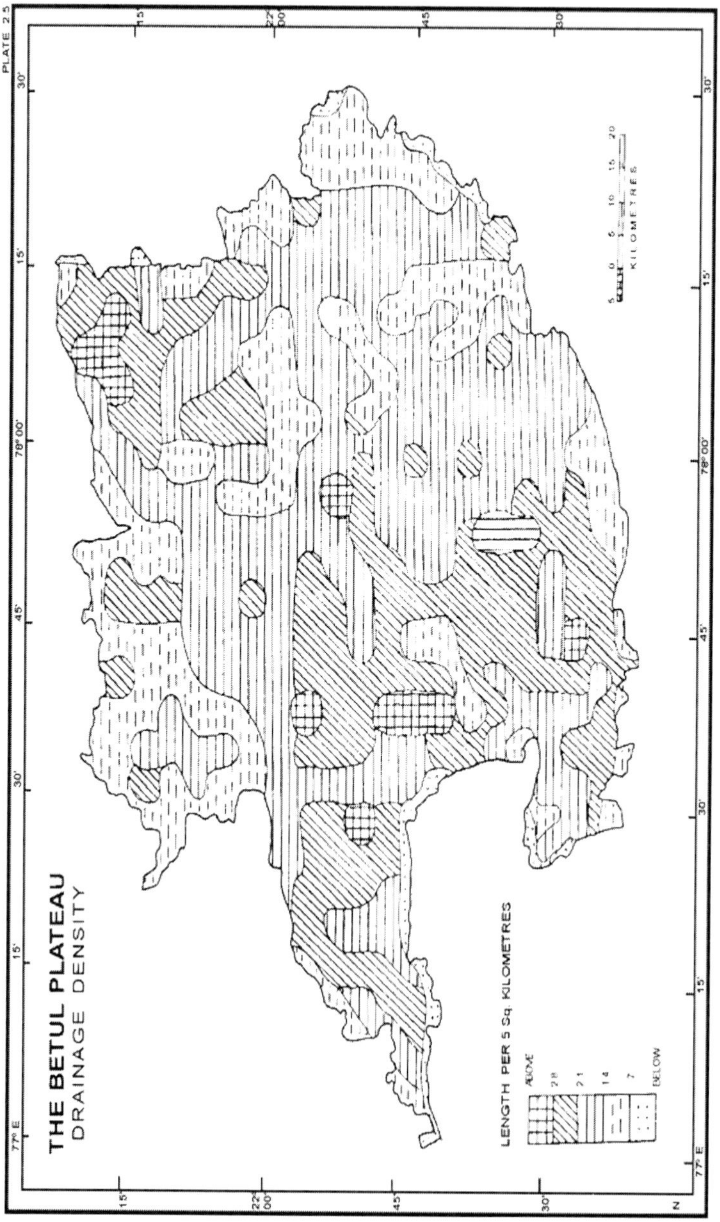
PLATE 2.5
THE BETUL PLATEAU
DRAINAGE DENSITY
LENGTH PER 5 Sq. KILOMETRES
ABOVE
28
21
14
7
BELOW
KILOMETRES

In the region, frequency varies from 1 to 20 kilometres per 5 square kilometres which is markable in the Plate 2.6. There is a gradual increase in variation of frequency from north to south and north-east. Greater land part of the study area consist medium frequency. High frequency covers the southern Asir hills and the south part of Chicholi. The northwestern and southeastern parts of the study area are covered by low and very low frequency.

The thickness of forests corresponds with the high degree of drainage density and frequency. In the area under study, the north eastern, west, south and middle parts are the extensive forest areas, as these areas enjoying high density and frequency of drainage. The forests are distributed along the course of the Tawa system to the north-east, the Morand to the north-west, the Tapti to the south-west and the tributaries of Puma to the south in the study area. Thus, customarily the area is divided into forest divisions and ranges according to the drainage system.

The density and height of trees in river valleys are noticed very high. It is the highest on flat alluvial deposits and on the lower and humid stopes of sheltered valleys. The river valleys which are protected from the wind have been noticed with a high density of the plants. The Tawa valley, the Machna valley, the Morand valley and the Tapti valley of the study area are densely forested. These river valleys have precipices, so it is very difficult to human reach therefore, it is safe from deforestation.

Variation in density and frequency influences the type of vegetation. This may be because of availability of moisture content in the soil. In the study area, the north east and middle parts of high drainage density consist of moist mixed forests whereas southern and western parts are covered with dry mixed forests on the same density and frequency. The middle part of the plateau, southerneast and western part with medium density and frequency have been occupied by dry teak forests. The bordering southern part with very low density and frequency consists of Salai forests.

Drainage is a dominant factor in the study area. On the banks of river the plants are being destroyed due to gulley erosion for example, on the southern slope of Asir hills due to gulley erosion, forests are being destroyed, thus deforestation has allowed gulley erosion in other parts of study areas which has reached at an alarming position. On the basis of above discussion it can be concluded that the climate of the study area is humid due to drainage system which is essential for the growth of vegetation. Thus the growth of vegetation and drainage go hand in hand to a long way.

CLIMATE AND PLANT LIFE

Native vegetation is looked upon as the best expression of the totality of a climate, because the plants are capable of measuring all the integrated climatic elements. Since climatic factors exert mainly a regional influence on plant life, the differences in the behaviour of a plant or a group of plants over extensive areas, as in given state may be considered as due to differences in climatic conditions *(Klages 1958, 44)*. Climate of any region provides the ecological limitations in which all organisms have their limits of tolerance and beyond which they do not grow as well as do not develop from their *Climatic climax* succession point of view. *(Charan 1992, 25)*

With this synthesis the climate of the study area is segmented into different profile basing upon variety of climatic elements. On this part the vegetation climate is a prime consideration which has been discussed at large whereas the customary and traditional climate has been narrated in brief.

The physiography of the study unit has a deep impact on climate because of this fact the climate of the region is unlike its surrounding areas like Narmda Basin and the Barar plains. The dominating factor is Satpura Range which spreads over northern, southern and western boundaries. The north-eastern and south-western parts of the region are the upper most areas. The height of the plateau is varying from 500 metres to 750 metres above mean sea level in comparison to the Tawa & Tapti Basin. This feature exerts considerable effect on temperature and rainfall of the region.

The sensible climate is that this area is less warm in summer colder in winter and in rainy season it is more humid.

TEMPERATURE AND PLANT GROWTH

The Betul plateau lies within or about the tropical belt and is situated to the south of the northern tropic. This geographical situation has an important effect on temperature. The temperature remains pleasant though experienced dryness for a longer time of the year. The mean annual temperature of the study area is 24.5°C. The temperature increases from February. The mean daily temperature in March is 25.05°C. The hottest month of the year is May, with the mean daily temperature 32.5°C. In July the mean temperature of the area remains 25°C to 25.2°C due to cloudy weather. (Table 2.3)

In August there is no considerable variation in temperature but in September and October it increases gradually due to clear sky. From November to January there is a fall in temperature and it is minimum 10°C in the month of December. The minimum temperature is known to fall even further in exceptional circumstances of a cold wave resulting from western disturbances moving north-east across North India. The seasonal changes of temperature are as follows:

1. Rainy season is more or less equitable and the post rainy period is warm to hot. Cold season is usually severe, while hot season is comparatively mild.
2. The summer heat on the plateau is mitigated by the altitude, which also ensures a cold winter. Temperature varies according to altitude, aspects and configuration of the ground. Thus, it is more equitable in plateau than in valleys.

   In August there is no considerable variation in temperature but in September and October it increases gradually due to clear sky. From November to January there is a fall in temperature and it is minimum 10°C in the month of December. The minimum temperature is known to fall even further in exceptional circumstances of a cold wave resulting from western disturbances moving north-east across North India. The seasonal changes of temperature are as follows:
3. Rainy season is more or less equitable and the post rainy period is warm to hot. Cold season is usually severe, while hot season is comparatively mild.

Table 2.3

Betul Plateau: Temperature of Betul

(In °C)

Month	Daily Maximum	Daily Minimum	Mean Monthly
January	26.40	11.1 1	08.75
February	29.80	12.50	21.50
March	33.60	16.50	25.50
April	37.00	21.10	29.05
May	39.30	24.80	32.05
June	35.00	24.40	29.70
July	28.10	22.30	25.20
August	26.90	21.70	24.30
September	28.30	21.00	24.60
October	29.30	17.20	23.25
November	27.70	12.00	19.85
December	27.00	10.30	18.65
Annual Average	30.70	17.91	24.37

Source: Climatological Table of observatories in India (1931-1961) Government of India, p-397.

4. The summer heat on the plateau is mitigated by the altitude, which also ensures a cold winter. Temperature varies according to altitude, aspects and configuration of the ground. Thus, it is more equitable in plateau than in valleys.

March of Temperature

1. **The Daily Range of Temperature** : Table 2.4 shows the daily range of temperature in the area under study which is the highest in the month of February (17.3°C) due to low humidity resulted from vernilization of leaves. This area is richly provided with vegetation which inhibits solar radiation

Table 2.4

Betul Plateau: Range of Temperature

(In °C)

Month	Jan	Feb	Mar	Apr	May	Jun	Jul	Aug	Sep	Oct	Nov	Dec
Temp. In '°C'	15.3	17.3	17.1	15.9	14.5	10.6	05.8	05.2	07.3	12.1	15.7	16.7

Source: Compiled on the basis of the Table 2.3.

to reach it, thus causes an increase in humidity. In other way, we can say that the daily temperature is influenced by vegetation indirectly. An increase in both mean daily minimum and daily maximum temperature alarms the plants to their preparation for scanty supply of moisture because of more evaporation.

2. **Monthly Range of Temperature** : From Table 2.5, it can be seen that the monthly range of temperature is highest in the month of May (41.7°C) and lowest in the month of January (5°C). The hottest month is usually May but occasionally it may be June, if the onset of monsoon is delayed. The coldest month is January but it can also be December and occasionally February. The combined affect of temperature is marked distinctive in different parts of region. The northern part experiences moderate, the middle part remains cold while the south-western part is marked semi hot.

The sheltered vallies of Tawa and Machna have low temperature thus consist of moist deciduous forests. The Mahadev hills towards the north-east have sub-tropical wet hill forests due to high temperature.

Each species of plants has its own maximum and minimum temperature beyond which their life activity ceases *(Kochhar 1967, 10)*. It is known that the temperature exceeding 48.9°C has inimical effect on plant growth. In the study area the highest maximum temperature is 43°C which never exceeds the critical limit and is therefore not a limiting factor. So the plant growth is safe from this

Table 2.5

Betul Plateau: Highest & Lowest Temperature for Betul

(In °C)

Month	Highest in the month	Lowest in the month	Mean
January	29.30	05.00	17.15
February	34.00	06.40	20.20
March	37.60	10.90	24.25
April	40.30	16.40	28.35
May	41.70	20.60	31.15
June	39.70	21.10	30.40
July	32.90	20.60	26.75
August	30.20	19.60	24.90
September	31.00	18.80	24.90
October	31.70	11.20	21.45
November	30.20	07.20	18.70
December	29.50	05.50	17.50
Annual Average	34.01	13.61	23.81

Source: Climatological Table of observatories in India (1931-1961) Government of India, p-3l 7.

damaging effect. Also plant growth remains dormant for the lowest minimum temperature up to 5.6°C. The temperature below the critical limit of 5.6°C is generally operative during the months of December and January and may extend from November and continue up to February also, thus restricting plant growth. The months of October and November and February to March are the periods of partial plant growth. During these months the mean minimum temperature does not rise above 18°C. Months of April to September are of maximum plant activity.

RAINFALL AND PLANT GROWTH

Rainfall variation is the most important ecological factor in determining the distribution of plants. The intensity, quantity and periodic distribution of rainfall are all important for vegetation *(Mehra & Khanna 1967-13)*. Rainfall data for six rainguage stations of the study area are available from metrological records. Fifty year data have been used for studying the characteristic of rainfall of the area under investigation. The seasonal and annual rainfall for various stations clears the following points:

1. Generally the monsoon sets in the second week of June and ceases sometimes between the last week of September and the second week of October. During the month of June to September the rainfall is about 85 per cent of total rainfall of the year.
2. The mean yearly rainfall for the study area is 1066.8mm whereas for Betul it is 1192.8mm for Multai 1011.4mm, Chicholi 1147.6 mm, Shahpur 1140mm, Betul rainguage station has maximum rainfall. It is located in the centre of study area.
3. The rainfall increases toward the north-east and central part of the area.

Distribution of Annual Rainfall

The area receives the mean annual rainfall of 1066.8mm. The central part receives the maximum rainfall (1192.8mm) while the south-east part receives the minimum rainfall (814mm). This indicates that the rainfall decreases towards the south-east and increases towards the north-east because of the Satpura range. The Satpura range receives much rainfall towards its northern part than the southern part.

The central part of the study area receives maximum rainfall including southern Bhanwargarh, Betul and central part of Amla forest range. These areas have moist deciduous forests. The Keelandeo hiils consist of sub-tropical wet forests due to heavy

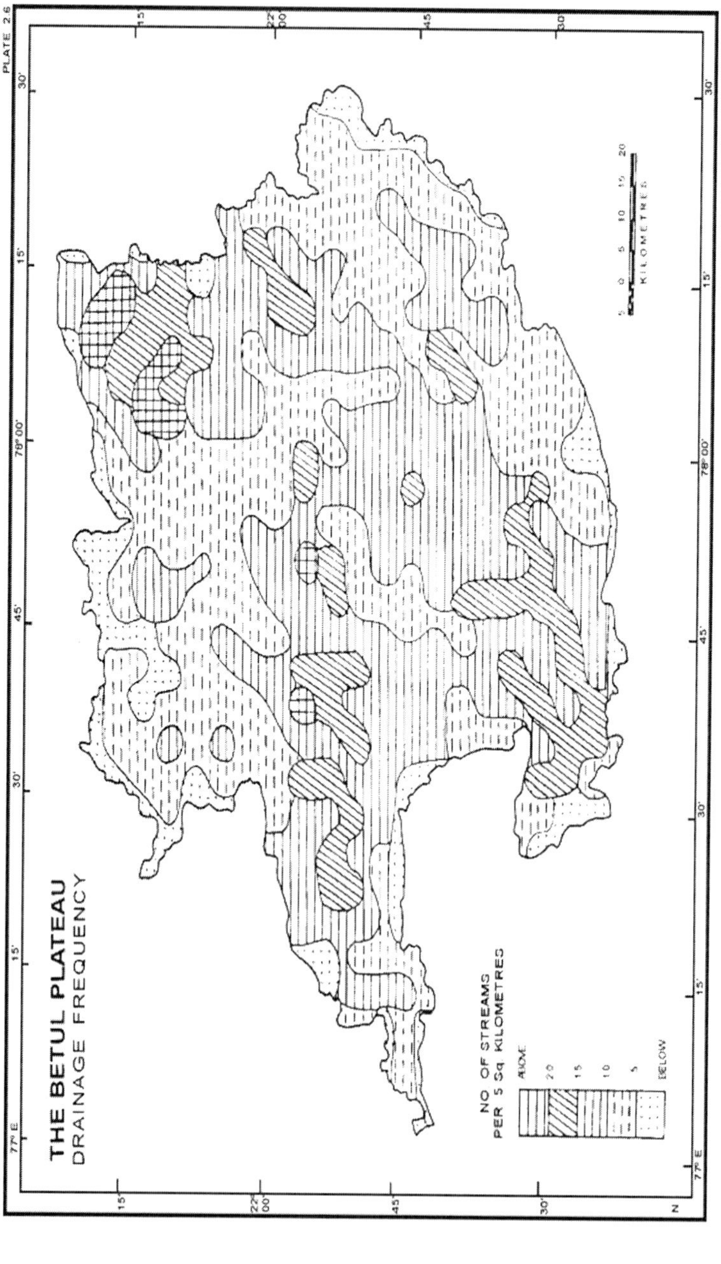
PLATE 2.6
THE BETUL PLATEAU
DRAINAGE FREQUENCY
NO OF STREAMS
PER 5 Sq KILOMETRES
ABOVE
20
15
10
5
BELOW
KILOMETRES

rainfall. A greater part of the plateau receives normal rainfall and is covered with dry deciduous forests.

Distribution of Seasonal Rainfall

Table 2.6 illustrates that the months of June to September are always rainy, thus constitutes a distinctly delimited aggressive rainy period. During this period the area receives 84.5 per cent of the total rainfall. The total mean rainfall during post rainy periods from October to February is 121.8mm. The months of March to May receive 4.2 per cent of the annual rainfall. These constitute the drought periods.

In the rains during rainy period surplus soil moisture favours the germination of plant species and good growth of plants. The post rainy season with partial rainfall i.e. whole of October is advantageous in prolonging the beneficial growth conditions of rainy season beyond that season during November to February though rains are received but in small quantity, it has a far reaching effect in improving the moisture conditions for the growth of plants. March to May constitute drought period during the period rains are received as light showers and are lost as a result of intense heat and when it occurs as concentrated showers it proves beneficial in lowering down the incidence of forest fire and causes the new flush of green grasses to appear.

Vegetation Climate

Vegetation climate is the climates within a few meter of soil surface, encomposing life sustaining zone suitable for certain range of plants. It is designed with the notion of relation of temperature and rainfall combinely and separately, in order to determine their deep impact upon vegetation. In the study area, month to month variation of vegetation climate is determined by preparing an Ombrothermic diagram *(Gaussan, 1960)*. The diagram (Plate, 2.7) is based on the Table 2.3 in which ombric curve remains above the thermic curve during June to October. This illustrates that the area receives maximum rainfall during this period. As the amount of available water is the most potent factor in determining the shape of the plants and their mode of life *(Hardy, 1913, 143)*. Therefore, these

Table 2.6

Betul Plateau: Distribution of Seasonal Rainfall

S.No.	Station	Rainy Season June to Sept. In millimeter	Winter Season Oct. to Feb. In millimeter	Summer Season Mar. to May. In millimeter
1.	Betul	1037.1	109.1	40.6
2.	Multai	0846.0	120.3	44.4
3.	Shahpur	1096.7	113.3	32.2
4.	Chicholi	0910.2	113.1	34.5
5.	Athnair	0658.0	124.0	31.3
6.	Bhainsdehi	0917.2	139.9	37.4
	Seasonal Mean	0925.3	121.8	46.7

Source: Climatological Table of observatories in India (1931-1961) Government of India, p-3l 7.

PLATE 2.7

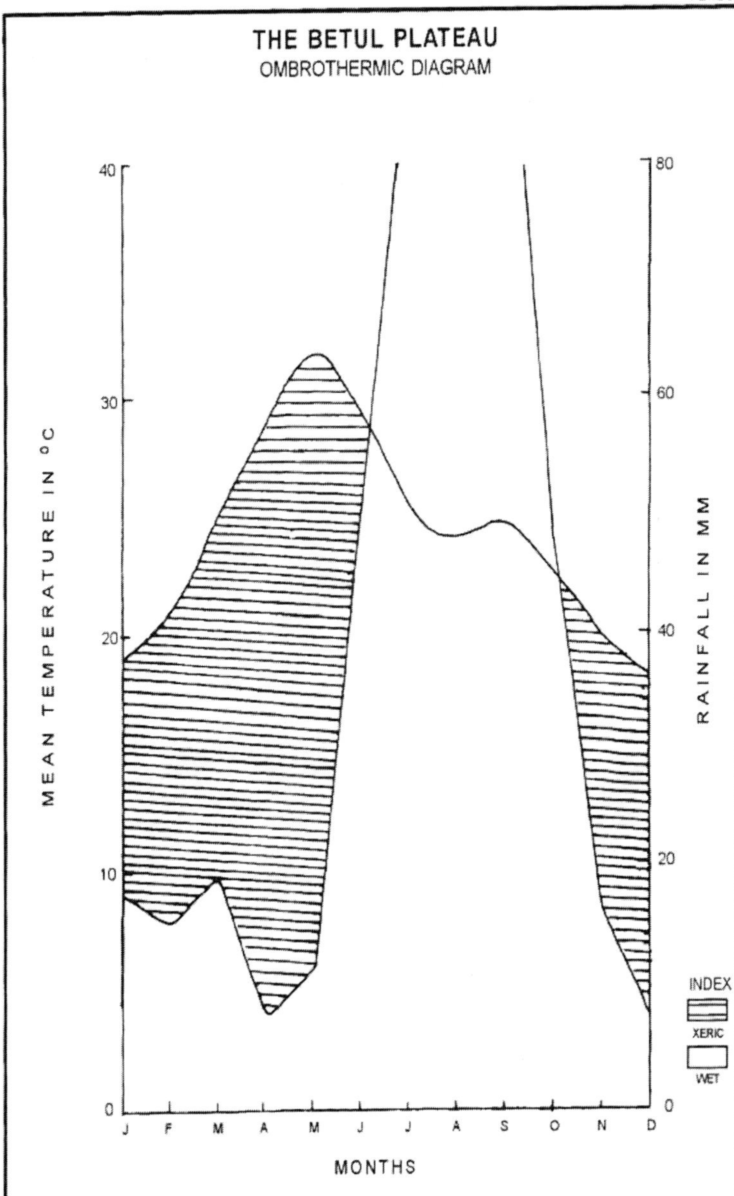
THE BETUL PLATEAU
OMBROTHERMIC DIAGRAM
MEAN TEMPERATURE IN °C
RAINFALL IN MM
40
30
20
10
0
80
60
40
20
0
J F M A M J J A S O N D
MONTHS
INDEX
XERIC
WET

months are significant by phenological point of view. During this period many of the herbaceous species attain their optimum vegetative growth. The moisture in soil so received favours the germination of various plant species. The seven months (November to May) are xeric, so the water balance remains unfavourable for the vegetation.

The formula proposed by Miller (1951), T/R is tried to determine the index of vegetation climate where T is mean annual temperature in °F and R is mean annual rainfall in inches. According to this formula the index of vegetation climate is 1.62 in the study area which represents humid vegetation climate. It is conducive to good growth of plants. The index of vegetation climate also shows regional variation due to variable temperature and rainfall (Table 2.7).

Table 2.7

Betul Plateau: Index of Vegetation Climate and Resulting Values

S.No.	Station	T/R index	Type Description
1.	Betul	1.61	Humid (i)
2.	Chicholi	1.67	Humid
3.	Shahpur	1.68	Humid
4.	Bhainsdehi	1.75	Humid
5.	Multai	1.90	Humid
6.	Athnair	2.36	Sub-Humid
	Seasonal Mean	925.3	121.8

Source: On the basis of data calculation (Table 2.3).

1. **Humid (Milder) Vegetation Climate Zone**: It covers the higher elevation of north-eastern, south-western Satpura range and sheltered valley areas (Plate 2.8). The zone is under moist deciduous forests.
2. **Humid Vegetation Climate Zone** : It occupies the northern and western foot hills and Betul-Chicholi plateau. In general, the vegetation is mainly consisting of dry deciduous type and the percentage area of forests is high in this zone.

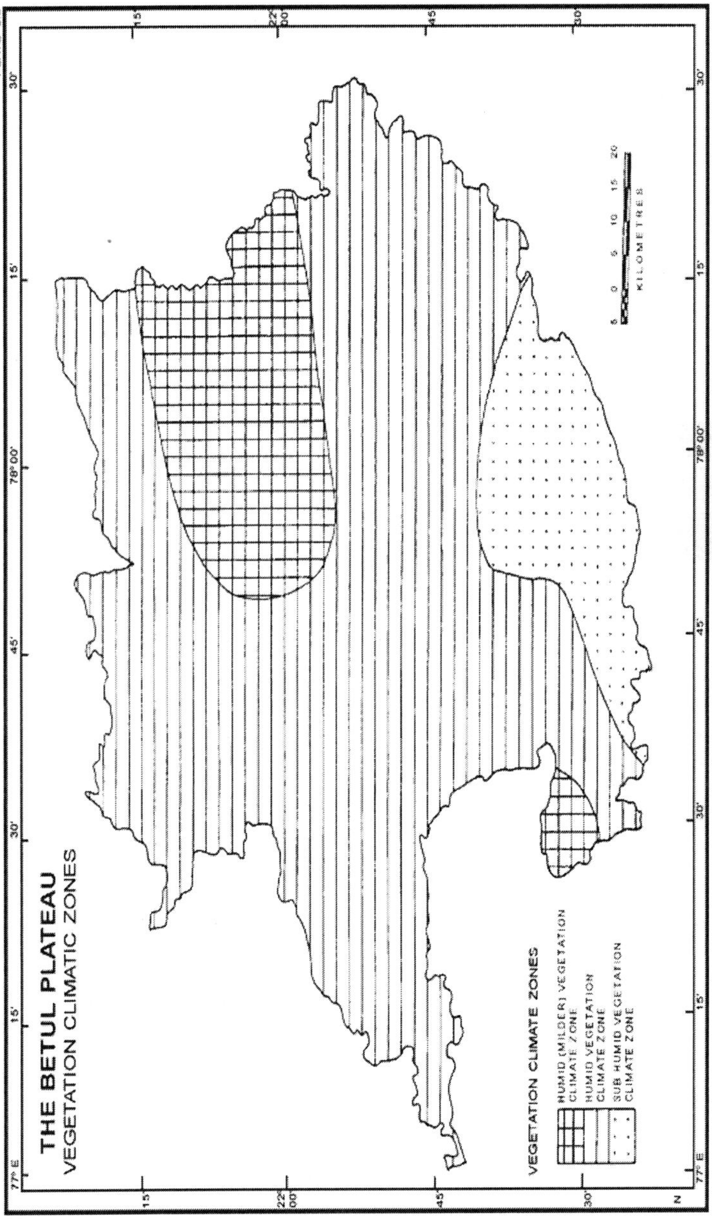

Source : Forest Department Records

3. **Sub-Humid Vegetation Climate Zone** : The southern scarpment and undulating Bhainsdehi plateau fall under this zone. It is characterised by Savanna type vegetation.

EDAPHIC PROFILE AND PLANT ADAPTATION

The soil is in fact the very hearts of the life layer known as the biosphere because it represents a zone, where in plant nutrients are produced, held, maintained and are made available to plants through their roots. The soil is a natural body of vegetation, mineral and organic constituents *(Joffe, 1949)*. The relationship of plant soil is an intimate one and the quality and quantity of plant product reflects the degree of soil fertility *(Greulacj et. al. 1962, 18)*. This is the soil layer where organic materials derived from plants and animals and mineral derived from the parent rocks are disintegrated and decomposed and are changed into elements.

Soils have been considered as the best indicator of geology, natural vegetation, and climate of a region and therefore represent the personality of the region in perfection *(Rawat and Pande 1989, 134)*. In the study area different types of geological formation occur which have produced variety of soil. The soils differ considerably within short distances as the topography shows sharps irregularities in the slope of land.

Soil Type

The soil of the region is largely developed in the situ and therefore, their characteristics are governed mainly by the nature of parent rock and nature of terrain. Major constituent rocks in the region are sandstones, shales, limestones, conglomerates, quartzite, gneiss and granite. Different combinations of these rocks have given birth to different types of soils which are further diversified by the nature of terrain (Plate 2.9). They vary from clayey to silty loam and sandy in texture, deep black red in colour.

On the basis of colour, depth and fertility status, soil of this region are classed into six soil types which can be put into three major groups of Indian soil classification scheme *(Natmo 1980, Plate 9)*.

PLATE 2.9

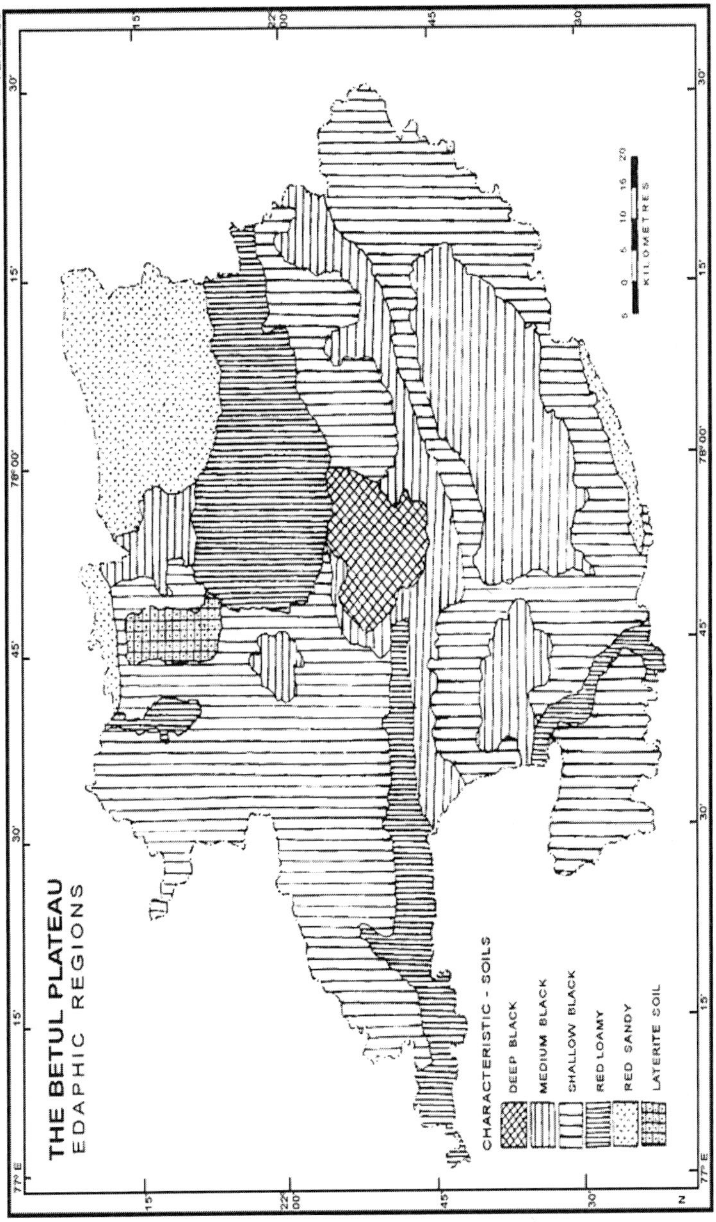

Source : Agricultural Atlas of India (Central India Plate 8-9)

1. Black Soil (Vertisols)
   i. Deep black soils
   ii. Medium black soils
   iii. Shallow black soils
2. Red Soils (Alfisols)
   i. Red sandy soils
   ii. Red loamy soils
3. Lateritic Soils

1. **Black Soil** : Black soil is derived from the Deccan Trap which occupies the major portion of the study area. However, they have also developed over sand stone shells that have more cohesiveness and cracks to a greater extent. The trap rocks though uniform in composition vary in the size of crystals and their resistance towards weathering. The depth of soil varies with its position along the slopes, presence of vegetation as soil cover, as well as the amount of rainfall. The soils are thus shallow in the drier parts as well as on slopes with more vegetation where considerable soil is lost through erosion, leading to no increase in depth of the soil. Following are the subtypes of the black soil.

   (i) *Deep Black Soil* : In the study area this soil is found in the cultivated tract around Betul town only. This soil is forming a triangle with vertices in small area at Betul, Betul Bazar and Khedi. The soil is deep enough and clayey in nature. These soils are well endowed with major nutrients viz, nitrogen, phosphorous and organic matter. These are comparatively fertile soil and therefore are under cultivation.

   (ii) *Medium Black Soil* : These soils are found on undulating plateau areas and also over smaller level tracts. Extensive tracts of this soil are the Bel River near Amla. The significant areas of this soil are the agricultural land along the road connecting Masood, Athnair, Gudgaon and Bhainsdehi from east to west. It is also found in lower Tawa basin. The physical character of the soil varies, it is black clayey loam on the low lying lands and fertile sandy loam on the slopes. Fertility of the soil is not high. It has sufficient minerals to support a fairly good forest crop but clearing this land and bringing it under cultivation, the forest and other natural vegetative covers are very limited in extent. The soil is found in small patches on the lower slopes of forested area where Bamboo forests grow as main crop.

(iii) *Shallow Black Soil* : These soils occur in high plateau area and hilly parts of the study region. The southern, western and northern parts of the region are shallow soil areas. The high Khamla plateau and Gawaligarh hills to the south, high Chicholi plateau and Kalibheet hills to the west and northern Bhanwargarh hills are characterised by these soils. Most often these soils are of residual type and sandy in texture. The depth is varying according to the slope. These soils are developed on sand stones and lime stones. Sand stones disintegrate into porous soils of low fertility. These soils are poor in mineral so are not suitable for farming and therefore they are left to support natural vegetation. The belt of dry type teak forests is extended over these soils. Salai forests are also scattered over these soils particularly on ridges and steep slopes where soils are thin layered.

2. **Red Soil** : Red soils cover a major part of land in the study area. They are confined to middle and southern parts bordering the deep black soil zone. They develop on sand stones. On the basis of texture and fertility status red soil can be put into the subgroups as follows:

(i) *Red Sandy Soil* : It is generally red gravel thickly strewn with reddish stones of a fair size but at some places stones are yellow fants. They are developed on the sandstones of the Jabalpur series and of the upper Gondwana System. The Jabalpur countries of the northwestern parts possess this type of soil. On northern Morand plateau, it is found as extensive belt while on southern scarpment it is found in small patches. These soils favour the growth of miscellaneous species. Some of the Barakar soils of this series are more clayey and support the growth of better quality of mixed forests.

(ii) Red Loamy Soil : These are formed of metamorphic crystalline rocks. The soil formed is of fair quality and depth. These soils occur in the comparatively low lying areas. The ravines of Tawa, Tapir and Machna are characterised by the presence of these soils. These are significant forest areas of the region. These soils are light and sandy on adjoining ridges and hill tops like southern Bhanwargarh and Badnoor range. However, the most important forest areas exist on the less acidic gneissis of these soils. These soils with greater moisture retentivity are most suitable for the growth of moist type of vegetation.

3. **Laterite Soil :** Formations of Laterite soils are high on flat ground. In the study area, these soils are found on undulating land part, hills to the north and southern Bhainsdehi plateau. These soils show considerable leaching and washing out the nutrients so they are less fertile. Due to their low fertility, these are not put under cultivation and are not adequately covered by vegetation. Thus creating large blanks and under stocked areas.

Revenue Based Classification

According to the latest settlement scheme, soil in the study area is also classified on the basis of the nature and productivity of different soils, rather than on their chemical composition which has adopted local and vernacular names. In this system of classification, soils have been classified into Kali, Morand I, Morand II, Morand III, Mutbarra I, Mutbarra II, Sihar I, Sihar II, Bardi I, Bardi II, Retari, Bari and Gairmumkin. A brief account is tabled as below: (Table 2.8).

Table 2.8

Betul Plateau: Tehsil-Wise Area under the Soil

(In Hectares)

Type of Soil	Betul	Multai	Bhainsdehi	Betul District
Kali	2,219	—	—	2,219
Morand	79,453	69,143	35,709	184,305
Mutbarra	28,293	51,916	46,954	127,163
Sihar	58,657	9,883	1,490	70,030
Bardi	149,362	222,060	179,011	550,433
Retari	31,318	45,442	6,395	83,155
Others	3,619	15,026	9,424	28,069
Total	352,921	413,470	278,983	1,045,374

Source: Furnished by Collector Betul vide his memo no. viii LR/64 March 1964.

It can be concluded that soils play an important role in distribution, growth and composition of growing stock. Nature has bestowed study area with much diversified soil. The region is predominantly occupied by the alluvial soil, sandy soil, red soil and black soil. These soil groups vary in their physical as well as chemical properties and eventually affect the plant growth. The black soils contain sufficient minerals to support a fairly good forest crop, particularly teak forests. Sandy soil is the most suitable for mixed type of forests. Vegetation cover with good fertile soil has been removed from area of steep slopes. So that the phenomenon has caused serious problem of soil erosion.

Soil Erosion

Soil erosion varies in intensity all over the study area. In this region soil erosion causes due to vulnerability, presence of steep slopes (30-40 per cent) and presence of uncovered hill tops. Gully and sheet erosion do occur, predominantly on steeper slopes. In the valleys of Tawa, Morand and its tributaries the erosion can be seen by rills and gullies. In the southern and central region gully erosion occurs due to the formation of ravines of the Tapti, Puma, Wardha and their tributaries. The land slides of sloppy lands due to heavy precipitation also cause erosion in southern region. In north-eastern part of the study area, gully erosion causing the maximum intensity of erosion in the Asir range. In this part land slides also occur during rains along Sarni to Ladi hill road.

In the study area soil is gradually displaced from its natural profile which is resulting in harmful effect, such as decrease in soil depth, water retaining capacity and minerals. Soil, air and moisture balance is also disturbed because soil aggregates are broken down. The micro-organisms perish due to erosion thus affect the humification and mineralization. The quality of organic matter in soil goes on decreasing. In plain areas, specially along the river tract, there arises a water logging problem which in tern results in alkalinity and salinity of soil. Because of above discussed effects the ground flora is greatly influenced and leading to the stage of extinction which is most important, as most of the medicinal plants belong to this group. Gully erosion reduces the forest areas of the total forested areas.

SOCIO-ENVIRONMENT

In the perspective of natural forces, factors and all other components affects man's multifarious and multidimensional activities along with plant and animal kingdoms, is understood as a complete concept of environment. It is needless to mention that man is a most active geographical factor and at the same time is also an alive agent of biosphere and thus intimately co-exists with the plants. Plants exert deep and direct influence on the mode of human life in many ways. Man, since early time has been getting food, fodder and fuel from the forests on one hand and on the other hand he is also getting such plants which cure his ailments Unfortunately such plants of medicinal value are depleting because of the rapid growth of population which result in a great paucity of agricultural land and hence man is trying to fetch agricultural land from forests by removing plant cover.

In the study area, a large portion of the population is tribal and hence ruthless cutting of the forest flora for domestic, industrial and other purposes is continued. It is pertinent to mention that these factors influence the distributional patterns of plants in many ways and therefore a stratified active participation of mankind is designed for the present study in this chapter.

Men with its surroundings make his environment and ultimately this is called the population of the area. The total population of the study area is 13,94,421 persons (2001 Census), which is 2.31 per cent of the total population of the state. Out of the sum 11,35,542 persons live in rural and 2,58,879 persons live in urban areas of the study region. The tribal population in the study area is 5,22,989 persons of the total population. The rural-urban ratio workout is 81:19. According to 2001 census there are 965 females per 1000 males in the plateau and generally, 46.3 per cent joint families are found in the area under study. It is sparsely populated because of hilly topography and extensive forest cover.

Density

In the study area, the density of population is 139 persons per square ki1ometre as against the state average of 196 persons per square kilometre. The density varies tehsils wise in the area. The Betul tehsil is densely populated as compared to Multai and Bhainsdehi Tehsil. The Bhainsdehi tehsil has low density.

The population of Betul tehsil and Multai tehsil is 263 persons and 183 persons per square kilometre respectively with high density. In Bhainsdehi tehsil density is 143 persons per square kilometre, because of dissected land which do not provide hospitable environment for human habitation. In forested areas the density is very low say about 3-5 persons per square kilometre.

Population Growth

The progress of population growth during the last nine decades has been quite fluctuating in response to the varying intensity of natural calamities, economic, climatic and political vicissitude. The decade 1901-1911 showed an abnormal growth of population i.e.

35-64 per cent which was the highest but a fall in the population of the study area by 6.83 per cent was noted due to the devasting influences, epidemic and famine in 1921. The population of the study area during three decades of 1921-1951 shows fluctuation in the percentage of growth rate. The decade of 1951-1981 witnessed an increase in population. The growths are during decade 1981-1991 was 27.68 and in last decade 1991-2001 it was 18.02.

Rural and Urban Population

The rural population of the study area is 81 per cent and the urban population is 18.62 per cent Bhainsdehi tehsil is entirely rural as there is no town in this tehsil. In Multai tehsil the rural population is 87.9 per cent and urban population is 12.10 per cent. Betul tehsil consist the ratio of 65 and 35 per cent of rural and urban population respectively.

Sex Ratio and Literacy

Males out number females in the study area with 7,09,525 males as against 6,84,896 females. Thus the sex ratio i.e. the number of males per 1000 females works out to 965 which are greater than the state average of 920.

The percentage of literates top total population in the area under study has gone up to 66.87 per cent in 2001 of which male literacy is 77.31 per cent while female literacy is 56.05 per cent. Due to inadequate educational facilities and lack of awareness towards education, the percentage of literacy in study area is low.

Occupational Structure

In the study area about 58 per cent of total population is living below poverty line. Due to lack of industrial development and literacy most of the inhabitants of this area depend on their traditional occupation such as agriculture, animal husbandry, forestry, household industry etc. The percentage of workers to total population is 48.4 of which 51.3 per cent are cultivators and 28.6 per cent are agricultural labour. The percentage of house hold industry and other workers is 2.52 and 18.5 respectively.

Tribal Habitat

The Betul plateau has significant concentration of tribal population sharing 38 per cent of the total population. The population growth of tribals increases four times greater during 1901-1991. The tribal population in Shahpur Tehsil is 63 per cent, in Bhainsdehi tehsil 61 per cent, 34.8 and 19 per cent in Betul and Multai tehsil respectively.

The Gond, Korku, Gauli and Kori are the primitive sub-tribals inhabiting the Satpura forest of study area. In fact the tribals are found at different level of cultural, economic organization depending on the nature of ecological isolation, cultural contact and resources of their habitat. They are inclined towards the natural resources for their subsistence and treatments. The use of plants or parts of it as medicine is common and popular in tribal areas and this is the reason that much first hand information has been collected from this area.

Tribal Economy

Tribal economy is characterised by land alienation, indebtedness, exploitation, labour system and poverty. Tribal economy is not agricultural, but forest based. Tribals have been totally dependent on forests for their survival so they readily destroy forests for small gains like fuel wood, food, shelter, financial needs etc. In the study area tribals earn money for various necessities by collecting and selling of forest produce. They collect various herbs, roots, rhizomes, tubers, flowers, fruits and seeds of many wild and cultivated plants growing in and near the forests. Tribals derive bulk of raw material for cottage industries from wild growing plants e.g. herbal drugs, fibers and flosses, cordage, mats and baskety, gum, reseins, musical instruments, katha, agricultural implements etc. The forests have suffered a tremendous loss due to tribal economy in study area which is responsible for disappearance of many plant species from natural forests.

Man and Environment

The story of man can be narrated in term of his changing

relationship with nature and forests. Man was depended directly on the natural product or on other animals that in their term depend on the natural product during early time. The important feature of this stage is that, man was not disturbing the ecology and was in the state of natural equilibrium. Gradually, as man has acquired greater skills in food gathering, his power of exploiting natural resources has increased. In the course of time his population began to increase faster that was difficult for the nature to support. Therefore, man began to select useful plants which would provide him food.

In the study area, as the pressure of population, grows, the encroachment over forest land have started by human being for cultivation. In this way clearing of forests, damage the forest eco-system. Agricultural statistic of the study area tells that during the period from 1971 to 1990, the overall forest area has decreased up to about 23,000 hectares. Another cause of forest destruction is increased live stock of people who live near forests. During 1985-86 total livestock was 7,36,600, but during 1991-92 it has reached up to 8,16,400. People utilize forest regions as pasture for their cattles ultimately over grazing exert direct influence on natural regeneration process of forests.

The herbivores graze on the seedlings of the trees and shrub species with the result that many herbaceous species, specially the rare ones from the medicinal point of view are seldom seen in all the stages of development. Urban development, commercial exploitation, irrigation projects, power plant i.e. Sarni thermal power plant, coal mines (Pathakheda), housing and illegal cutting of trees. These activities have affected the ecological balance affecting the plant life, a potential resource which is bound *(Kachroo, 1984, 5)* to play a significant role also in the forest degradation.

CONCLUSION

Each stage in the life cycle of a plant is greatly influenced by a number of environmental factors. Therefore in the study area, it has been observed that the environment in its diverse manifestations have offered different possibilities of varied flora in different areas. In this way it can be concluded that:

1. With regard to geology and vegetation altogether a different picture has emerged that the areas with trap formation have dense teak forests while, mixed forest have been occurred in the areas of metamorphic and sandstone rocks formation.
2. Varied topographic features of the area have resulted in the occurrence of various type of forest flora. The higher altitude is covered with wet tropical forests whereas lower humid slopes consist of moist forests. The plateau areas and flat hills are characterised by dry type of forests.
3. In regional personality of Betul plateau, drainage is the most dominant factor. In hilly and high plateau areas, rapid gulley erosion has resulted due to high drainage density and frequency which ultimately leads to the degradation of forest flora. On the contrary, lower plateau areas along with water bodies are covered with dense forests.
4. Due to humid vegetation climate, the area is covered with climatic climax type vegetation. The lower areas to the north eastern and middle portion support growth of moist forests because of more humid climate whereas dry forests are found in the semi-humid climate in the south part of the area.
5. Soil is also one of the dominant factors influencing the type and density of forests. Alluvial soil is characterised by luxuriant forests, while in shallow soil areas low plant density and dwarf plants are marked. Moreover, the black and sandy soil areas are covered with teak and mixed forests respectively whereas lateritic soil is characterised by Salai forests.
6. Because of human influences the forests of the region are confined only to the hilly and undulating plateau areas. The fertile black soil areas of forests have been subjected to an axe and plough since long back and because of this interference the forest cover is being depleting gradually.

REFERENCES

Charan, Anil K., 1992, *Plant Geography,* Jaipur, Rawat Publications.

Dikshit, K.R., 1984, *A Prologue to the Symposium on Geography and Teaching of Environment.* Deptt. Of Geography, Poona University, 59-66.

Gaussen, H., 1959, *The Vagetation Maps* Travaux do la Section Scientific Technique Tome I, Fascicule IV, Pondichery, Institute Francais.

Greulacj, Victor A and Edision, A.J. 1962, *Plants,* London, John Wiley and Sons.

Hardy, M.E., 1913, *An Introduction to Plant-Geography.* London, Oxford Press.

Joffe, M.E., 1949, *Pedology.* New Brunswick, Rutgers University.

Kachroo, P., 1984, *Floristic Sociology and Conservation of Plant resources of Pir Panjab Forest range.* MAB Project report, 1-14.

Klages, K.H.W., 1958, *Ecological Crop Geography.* New York, Mac Millan.

Kochher, P.L., 1967. *Plant Ecology Genetics and Evolution.* New Delhi, Atmaram and Sons.

Mehra, V.B., and Khanna, S.K., 1976, *Plant Ecology.* New Delhi, S.Chand & Company Publication.

Miller, A.A., 1951, *Transactions and Paper Institute of British Geographer,* no. 17.

Negi, S.S., 1986, *Geo-Botany of India.* Delhi, Periodical Expert Book Agency.

Rawat, D.S. and Pandey, R.K.., 1989, Soil and Minerals of Almora District. A Geographical Survey in, *Environmental Issues and Programmes;* Ed. By I.Mohan, New Delhi, Ashish Publishing House.

Sharma, A.K., 1988, Biotic Resources of the Ranthambhore National Park in *Environmental Resources the Crisis of Development.* Ed by H.S. Mathur, Jaipur R B S A, Publishers. 91-102.

Singh, R.L., 1971, *India: A Regional Geography,* Varanasi, National Geographical Society of India.

Singh, S. and Dubey, A., 1983, Environmental Management - Some

New Dimensions, in *Environmental Management,* Ed. by L.R. Singh, et. Al, Allahabad University, Geographical Society, Geog. Deptt., 72-86.

West, W.D., 1959, The Source of Deccan Trap Flows, *Jour. Geol. Soc. Ind.* Vol. I, 44-52.

3

Regional Flora

The preceding chapter dealt with the role of environment. A set of surrounding conditions constitute the habitat of the plant community. These environmental factors or habitat factors influence the morphology and physiology of plant communities. Because of spatial variation of environmental conditions, there is a markable variation in the composition, structure and size of the plant communities *(Tivy, 1982, 27)*. Thus different species of plant grow together, with different life forms. These communities of all types of vegetation are referred to as flora. The development of different species of regional flora takes place through the process of adaptation, competition and natural selection, thus ultimately known as indigenous plants.

The florastic distribution requires detailed study of the individual plant community which involves many concepts like frequency, dominance, constancy, presence and fidelity. The botanists study the flora by classifying it into classes, orders, alliances and associates but it is modified when the stands are treated as continum. But geographers prefer the ecological classification as ecology reflects the vegetation. The variation in ecological conditions causes direct or indirect effect on type and distribution of vegetation

(Botkin and Keller 1982, 506). The ecosystem itself manifests through the presence of predominant species in upper tier. The vegetative under growth also varies with ecosystem. The regional flora can be described by dividing regions into major forest types. Each forest type has been discussed along with the stratification and ecological status of plant species.

Thus, keeping the above view and Geographer's concept in the mind the study unit has been divided into forest types. Each forest type is tried to be discussed at large. The stratification and ecological status of forest types have been determined in order to give a clear picture of environmental impact on different plant communities and their distribution.

DISTRIBUTION OF FOREST

In the Betul plateau the total forest area is about 4042 sq. kilometres which covers 40.17 per cent of the study area. This percentage is much more than state percentage (32 per cent). The flora of the region is less disturbed because of the difficulties of the communication, natural restriction on the expansion of agriculture and the low density of population. The village areas are managed by the Revenue Department containing sufficient vegetation. The area under study may be divided into the northern, southern and western section. The hills and high plateau area of these sections are covered with the forests (Plate 3.1).

In the northern part of the area the forests are scattered over the Mahadev hills. In southern part they occupy the Gwaligarh hills. Like-wise in the western part, forests are found at the peak of eastern Kalibheet hills and high Chicholi plateau. The central and eastern parts of the Betul plateau are used for agriculture as they contain black soil and make a triangle. Betul is located at the top of this triangle with other unforested areas like Athnair, Multai, Amla, Gudgaon, Bhainsdehi and Chicholi. These are level tracts and used for agricultural purpose. The hills and dissected plateau have patches of black soil which support dense forests.

The northern part of the study area consists of 64 per cent forests of the total land. They are extended over a wide area of Asir

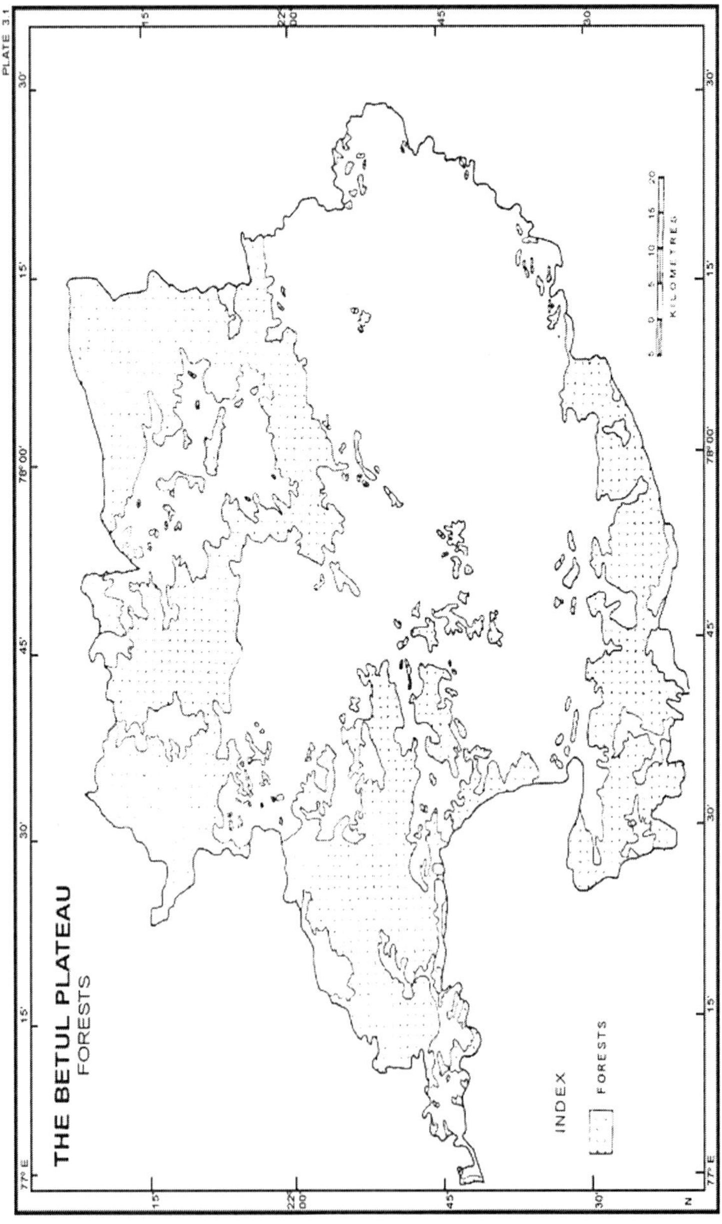
PLATE 3.1
THE BETUL PLATEAU
FORESTS
INDEX
FORESTS
KILOMETRES
5 0 5 10 15 20
77° E
15'
30'
45'
78°00'
22°00'
N

hills, Bhanwargarh hills and in the Machna valley. The Keelandeo hills to the north-east and the upper Tawa valley to the south of it consists of revinous forests. The residual hills of Bhanwargarh, Saoligarh, Chilpokna, Phoplia and the middle valleys are extending upto the northern boundary and the central part of the Morand valley are unforested areas so used for agriculture.

In the western part, the forests occupy 54 per cent land of the total Geographical area. The Tapasari hills, Alampur hills high plateau and western part of the Tapti valley fall under his forested area. The southern part of study area has 26 per cent forests of total land. The middle part of the Tapti valley comes under it. In this valley forests spread over its northern portion as well as southern portion.

To the south of it, forests cover the Neelgarh hills and Jhallar hills. Besides the Tapti valley, sourthern Khamla plateau, the peaks of marginal Gawaligarh hills, spur and the southern slopes of the plateau are also densely forested area.

Area under forests as reported by the land record at lower level than district does not include area of reserved and protected forests. Therefore, the following discussion on tehsil Level distinction of forests does not give the true picture of the area under forests in different tehsil of the region. The tehsil wise distribution of forest area shows that Shahpur tehsil is thickly forested with 58.8 per cent forest while Multai tehsil has only 18.9 per cent forest (Table 3.1)

Table 3.1

Betul Plateau: Tehsil-wise Forested Area

S.No.	Tehsil	Total Area (in hectares)	Total Forest Area (in hectares)	Percentage of the Forest Area
1.	Betul	2,79,731.67	1,13,762.00	40.60
2.	Shahpur	1,04,194.70	61,325.20	58.85
3.	Bhainsdehi	3,63,681.29	1,78,597.35	49.10
4.	Multai	2,57,098.80	48,710.00	18.90
	Total	10,04,706.47	4,02,394.55	40.05

Source: Complied from census of India, Village and Town Directory 1981, series for Betul District.

There are 1357 villages in the study area. The northern part consists of 40 per cent villages out of 72 per cent villages are forested while 28 per cent have no area under forest. In the southern part out of 60 per cent villages 52 per cent area is under forest while remaining 48 per cent is unforested. These calculations show that the northern part covers the largest area under forests than the southern part (Table 3.2).

Table 3.2

Betul Plateau : Village-wise Distribution of Forests with Percentage

Name	Total No. of Villagers	Number of Villages		Percentage of Village	
		Unforested	Forested	Unforested	Forested
Betul	549	154	395	28.50	71.90
Shahpur					
Bhainsdehi	381	93	288	24.40	75.60
Multai	427	295	132	69.08	60.05
Total	1357	542	815	39.94	60.05

Source: Complied from census of India 1981 series for Betul District.

The Tehsil-wise study shows that Bhainsdehi Tehsil has 75.6 per cent forested village. In Betul Tehsil, including Shahpur it is 72 per cent and in Multai Tehsil forested area is about 31 per cent. Thus we see that Bhainsdehi Tehsil is the lowest with minimum 24 per cent unforested villages and Multai Tehsil stands at the top with 69 per cent unforested area.

ADMINISTRATIVE PROFILE OF FORESTS

The management of forests in the study region is done by the Forest Department of Madhya Pradesh. The central circle of forest, under the forest department has divided the forests of the area into three divisions by the administrative and management point of view. These divisions are named as the North Division, the South Division and the West Division. The north-east part of the north division falls under the management of forest-corporation. This area is greatly eroded by the tributaries of the Tawa. A programme namely Rampur-Bhatodi Project has been launched in this area to check the soil erosion. The forest divisions are sub-divided into the forest ranges which are further divided into several series made of many blocks (Plate 3.2). The area under study has 15 forest ranges out of which five are included in north division, six in south division and four in the west division (Table 3.3).

Table 3.3

Betul Plateau : Range-Wise Forest Area

No.	Range	Total Geographical Area (in hect.)	Total Forested Area (in hect.)	Percentage of the Forested area
1.	Bhaura	31,640.0	29,416.5	92.90
2.	Gawasen	18,200.1	16,508.9	90.70
3.	Tawdi	36,876.9	25,045.0	67.91
4.	Mohda	49,744.0	32,049.0	64.42
5.	Barbatpur	41,360.2	25,704.1	62.14
6.	Athnair	45,083.0	27,394.0	60.76
7.	Sarni	54,061.0	29,926.6	55.36
8.	Sawalmendha	59,053.0	28,925.0	48.98
9.	Betul	49,080.2	21,659.7	44.13
10.	Chicholi	48,400.0	20,667.0	42.70
11.	Saoligarh	38,126.1	15,351.0	40.26
12.	Bhainsdehi	83,294.0	27,752.0	33.31
13.	Tapti	96,129.0	25,441.0	26.46
14.	Amla	91,849.0	23,801.0	25.91
15.	Multai	2,30,674.0	23,765.0	10.30
16.	Rampur-Bhatodi	32,377.5	30,729.5	94.91

Source: Circle Office of Forest, Betul

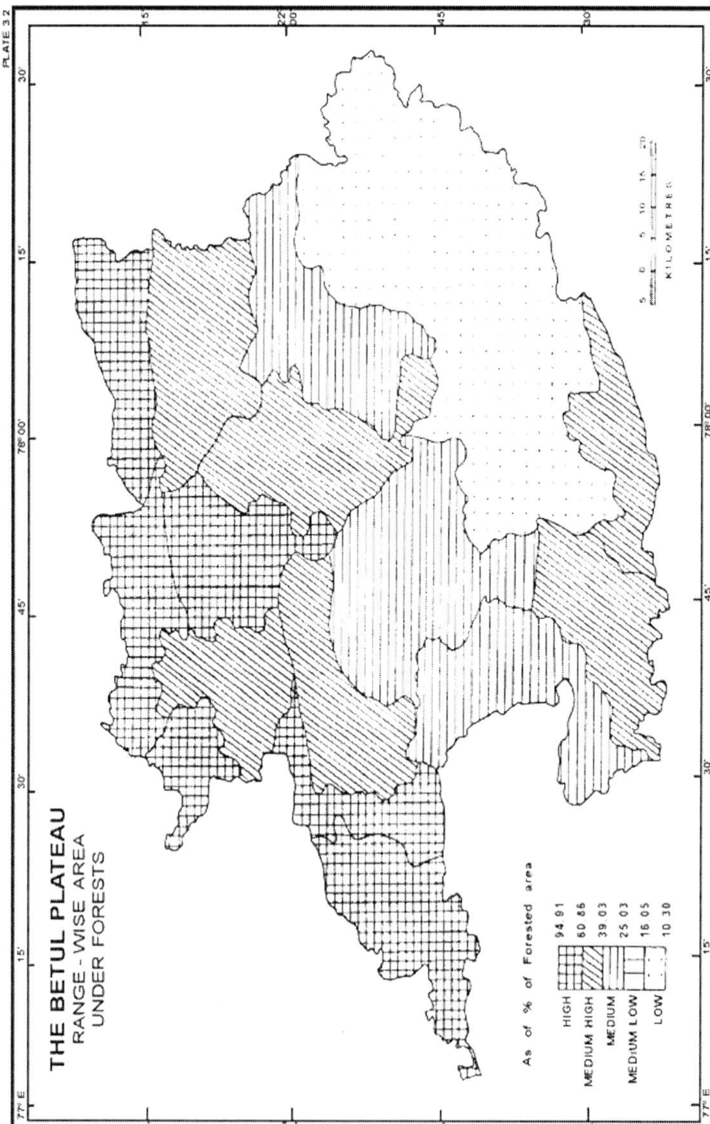

Source : Foreest Department Records

The forests have also been divided for administrative purposes into three classes viz; Reserved, Protected and other (undermarcated) forests (Plate 3.3). The area extent of these types of forests in different divisions is summarized in (Table 3.4).

North Division

The north division is spread over hilly and slopy land to the north direction of the study area. It has 29 per cent forest area of the total forests in the region of about 1,23,215.8 hectares. In this division 64 per cent area is under forest out of which 87 per cent area is of reserved forests and 13 per cent is of protected forests. The north hilly area of the division is covered with moist deciduous forests while the southern portions of the division have mixed and poor quality forests. The division consists of 50 per cent teak forests, 27 per cent mixed forests 13 per cent Salai and the remaining poor quality forests. The density of the trees is 0.5 to 0.9 in plain area due to favourable conditions. These conditions resulted in good and high quality of teak forests which resembles the teak of Burma.

South Division

Greatly dissected and undulated part of the plateau is included in the South Division with 25.9 per cent forest land out of total geographical area. In 58.6 per cent of the division reserved forests are spread while 20 per cent forests are demarcated. The more even part with trap soil is used for agriculture. This area is highly eroded by the Tapti, the Wardha and the Purna River and their tributaries, thus resulted in the formation of ravines. The ravines formation causes variation in types and density of forests. The density falls upto 0.5 in some parts. In this division about 42 per cent area is under teak forests, 13 per cent is under mixed forests and remaining area consist Salai and other poor quality forests.

The northern part of the division along the river Tapti is supplied with dense teak forests. The southern part, the Khamla plateau of the division is under mixed forests while bordering hilly southern slopes consists of Salai forests.

Table 4

Betul Plateau : Administrative Classification of Forests

(in hectares)

Name of the Forest Division	Total Geographical Area	Forest Area			Total Forest Area	Percentage of Forest Area in Division
		Reserved Forests	Protect Forests	Other Forests		
North Division	1,94,351.5	1,01,656.9	15,256.90	6,328.00	1,23,215.8	63.4
South Division	6,06,082.0	92,075.00	31,633.00	33,370.00	1,57,078.0	25.9
West Division	1,73,147.0	60,051.00	22,335.00	10,726.00	93,112.00	53.77
Rampur Bhatodi Project	32,377.50	30,184.00	5,455.00		30,729.50	94.9
Total	10,05,948.0	2,83,940.9	69,770.40	50,424.00	4,04,135.5	40.1

Source: Circle Office of Forests, Betul 1998.

PLATE 3.3

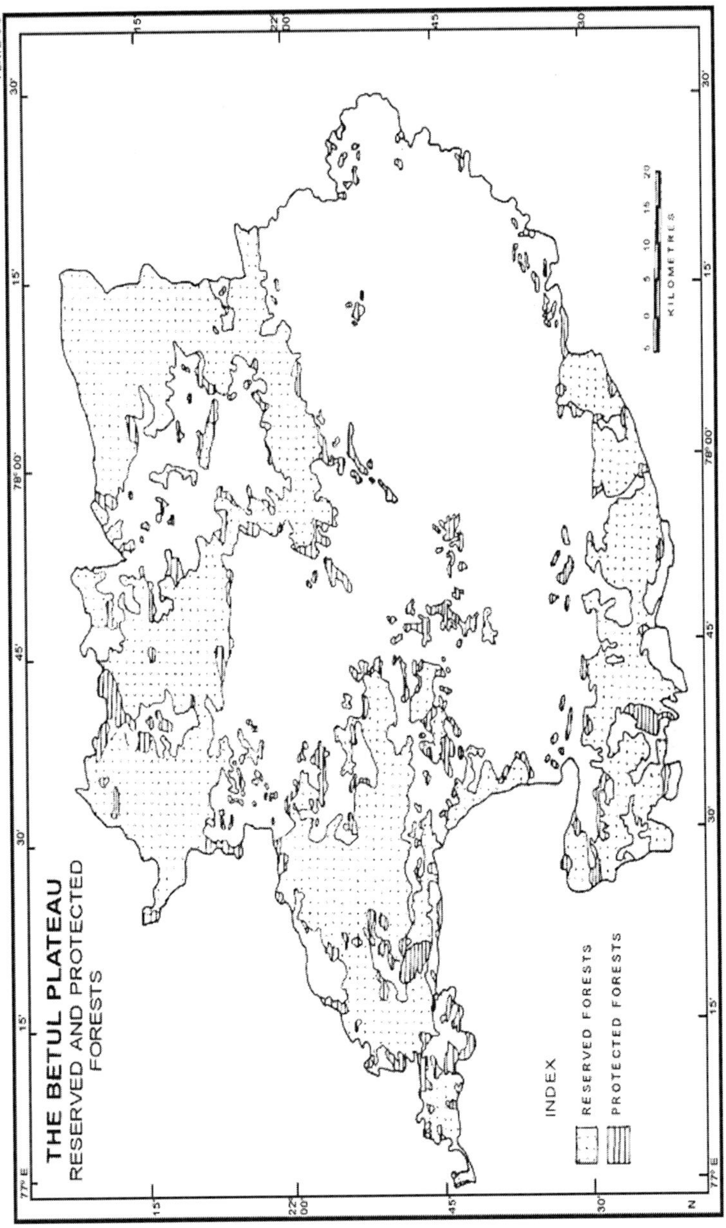
THE BETUL PLATEAU
RESERVED AND PROTECTED FORESTS
INDEX
RESERVED FORESTS
PROTECTED FORESTS
KILOMETRES
77° E
78° 00'
22° 00'

West Division

The west division is scattered over the hills of western part, high Chicholi plateau and undulated valley of the Tapti. This division occupies 25 per cent forest area of geographical area. The ratio of the Reserved, Protected and undemarcated forests in the division is 60.5 per cent, 24 per cent and 15 per cent respectively. The Tapti and its tributaries divide the division into small blocks. The Saoligarh hills and high Chicholi plateau in the northern part of the division have large blocks of teak and poor quality mixed forests. The river valleys and lower more even land of Chicholi plateau with black soil are used for agriculture. The division comprises different kinds of forests because of variation in natural conditions. In this area mainly poor quality teak forests occur but sometime Bamboo forests can also be seen in between them. The specific density of the crop is 0.6 to 0.7 but sometimes it decreases upto zero.

TYPES OF FORESTS

As per classification of forest types of India by Champion and Seth (1968), following types of forests are represented:

1. TYPE 3B : South Indian Tropical Moist Deciduous Forests.
   (i) Type 3B/C-1
   (ii) Type 3B/C-2
2. TYPE 5A: Southern Tropical Dry Deciduous Forests.
   (i) Type 5A/C-ib
   (ii) Type 5A/C-3
3. TYPE 4F/R-5-Riparian Fringing Forests.
4. TYPE 5/E-2-Boswellia Forests.

Type 3B

Tropical moist deciduous forests can be seen in the southern part of the study area which is divided into two groups. First type 3 B/C-1 South Indian Tropical moist decidous slightly moist teak forests occur mostly on metamorphic and crystalline and are confined to the valley of the Machna and Tawa river in Shahpur Tehsil due to alluvial soil and high moisture content. These forests are confined to the Baretha Ghat of Betul range, the eastern part of the Sarni range in Tawa valley, humid slopes of Nanda and Godhakhar of Tapti range and humid slopes of sheltered valley in Tapti range. The other type 3B/C-2 mixed forests occur on metamorphic and are confined to high plateau characterized by the presence of large proportion of evergreen plant species. These trees occupy the higher elevation of Keelandeo and the middle part of the Amla range.

Type 5A

Southern Tropical dry deciduous forests include most of the forest in the locality. They are classified into two groups. First group consist 5 A/C ib. Southern Tropical dry deciduous teak forests. On trap formation they cover a part of Gawasen and Bhaura Ranges and over crystalline rocks in part of Gawasen Range. These forests occur in northern part of Saoligarh range, the northern and western part of the Tapti range, the eastern part of Dabka range and Sawalmendha range. The teak forests on higher elevation in Betul and Amla ranges and on poor soils in Asir range also fall in this type. Various factors such as medium to deficient rainfall, a wide range of altitude and greatly varying condition cause occurrence of these forests. The second group includes 5 A/C -3. Southern tropical dry deciduous mixed forests in a small area near Handipani forests village in Bhaura range and areas near Sarni range, southern part of the Amla, the whole of the Asir range, due to sandstone, Khamla plateau, area of Bhainsdehi range and southern part of Athnair range. The main cause of occurrence of these forests is the presence of schists rocks just below the surface of water logging during the rains.

Type 4F/R-5

Riparian fringing forests are confined to lower banks of Machna river in hilly portion on made up soil along water courses and are characterized by the presence of very few species like Kohu, Gular etc. attaining large size and standing wide a part with smaller trees and shrubs in between.

Type 5/E-2 (Boswellia Forests)

Though Salai occurs all over in small patches on ridges and flat top in type 3 B/C-1(C)and 5 A/C-1(6), Salai forests occur in parts of Gawasen range with occasional portion of teak and absence of bamboo. The southern part of the study area viz; Dabka range, Sawalmendha and Tapti range are characterized by the presence of Salai, preponderance of Salai occurs on southern aspect, shallow soil, on rocky slopes or dry tops of hills.

VEGETATIONAL DIVISION

We know that adaptation is the result of prevailing ecophysiographical conditions already which were established from thousand years ago over the Satpura. To know the actual position of correlation in between the present ecophysiographical conditions and phytogeographical adaptations which have already been taken place in the plant species, the auther made some important field tour programmes on feet in the different part of Betul plateau.

The physiographic factor which are directly or indirectly influencing the phenomenon of adaptation in the body of plant species viz; soil moisture, distributional position and pattern of particular plant species, type of habitats. From the floristic point of view, existing edaphic factor plays a dominant role after the predominating primary role of prevailing climatic conditions throughout the area under study. Records of soil moisture percentage showed that it gradually increases from the vegetation under basin and compact soil areas while hilly and rocky areas vegetation which abviously coincide with the decrease by soil moisture. On the basis of field investigations it can be said that the distribution of it species is checked primarily by their physiographic formation and secondarily by the climatic limitations. Thus the condition within the division of vegetational associations of each vegetation division approximately coincides with the edaphic characteristics of its soil division. The area under Betul plateau has been studied in six vegetational division type as described below :

1. **Tectona-Anogeissus type** : Approximately 11 per cent of total forests area of study region is covered with this kind of

vegetation division. The division falls under the central highland and is observed in a scattered manner. The locality includes 5 study sites viz; The Machna, Tawa, Morand valley, southern part of Bhanwargarh hills and the area round about Nanda-Gadhakhar of Chicholi plateau. Altitudinally the whole locality is situated on gentle slopes with sheltered valley and usually northern aspects. The division is broadly divided into three major physiographical formations:

(i) Narrow alluvial plains which approximately cover 35 per cent area.

(ii) Lower and humid slope which covered about 45 per cent area.

(iii) Conglomerate area of Morand and Tawa valley which covered approximately 20 per cent area of the division.

Two important factors responsible for the occurrence of this type of vegetation are rainfall above 80 cm. and the nature of the underlying rocks.

This vegetation is confined to the area of soils derived from crystalline rocks. Crystalline rocks disintegrate into loamy or clayey loam soils. However, the more important forest areas exist on the less acidic gneisses and in the soil formed of fair quality and depth in valley, and the moisture regime is typie ustic. The mean annual soil temperature is more than 21°c qualifies the area for *Hypothermic* soil temperature regime. Where the accumulation of disintegrated rocks is thicker and well drained with acidic pH range of 6.5 to 7.0. Soils are low to medium in N and P, medium to high K contest. About 50 to 75 per cent soils are deficient in Zn and S. Nutrients like Fe and Mn are sufficient in these soils.

Trees are cylindrical and well formed with a good bole formation is observed and it is one of the important characters among the plant species of this vegetation division. The top canopy is almost entirely composed of teak (Tectona grandis). Individual tree may attain a height of up to 30 metres. The distribution of plant species varies according to the nature and distribution of physiographical formation and existing habitats. This type of vegetation has main vegetation group approximately 62 per cent trees, 17 per cent shrubs, 9 per cent climbers, 10 per cent herbs and 3 per cent grasses (Table 3.5). Thus, the plant species as per Raunkiaer's life-form classification are approximately 42 per cent *Mesophanerophytes* trees, and 52 per cent *Microphanerophytes* trees. Life-form spectrum of the plant species for this division shows that the plant species are *Phanerophytic* in nature.

Main associates of teak in the top canopy include Terminalia alata(Saja), Lagerstromio parviflora(Lendia), Anogeissus latifolia(Dhauda), Adina cordifolia(Haldu), Ougenia oojeinensis(Tinsa), Madhuca longifolia (Mahua), Mitragyne parvifolia(Kalamb),Miliusa tomentosa (Kari), are abundant in this vegetation division. The middle storey is often composed of Emblica officinalis (Aonla), Flacourtia ramentchi (Kakai), Buchanania lanzan(Achar),Semecarpus anacardium(Bhelwan), Butea monosperma (Palas), Dondrocalamus strictus(Bamboo) etc.

Table 3.5

Betul Plateau : Vegetation Groups in Various Vegetation Division

S.No	Vegetation Division	Trees	Shrubs	Climbers	Herbs	Grasses
1.	Tactona-Anogeissus Type	61	17	9	10	3
2.	Anogeissus-Terminalia Type	53	21	7	15	4
3.	Tactona-Lagerstroemia Type	43	17	7	19	14
4.	Chlorxylon-Diaspyros Type	48	19	8	18	7
5.	Boswellia-Sterculia Type	29	23	7	29	12
6.	Arjuna-Ficus Glomerata Type	23	19	8	34	16

Source: Based on Field Work.

Common shrubs found in the under growth are Nyctanthes arbortristic (Siharu), Helicteres isora (Marorphal), Woodfordia fruticosa (Dhabai), Indigofera eassiodes (Neel),Grewia hirusta (Khareta), Lantana camara (Lantana). Bauhinia vahlii (Mahul), Butea superb (Palasbel), Abrus precatoricus (Ghungchi), Dioscorea bulbifera (Ratalu) are the common climbers of the division.

Among sedges and grasses, Achyranthus aspera (Apamarg), Cassia tora(Panwar), Mimosa pudica (Lajwanti), Swertia anguslifolia (Chiratta) herbs are dominantly as well as abundantly distributed and frequently formed patches of pure population. Some of the dominant grasses are Themeda quadrivalvis (Gunari), Cynodon dectylon (Dub), Vetiversia zizanioides(Khas), Apluda mutica(Ponai) and Ischaemum laxum(Sainar).

2. **Anogeissus-Terminalia type**: This vegetation covers approximately 9 per cent of the total area under forests. The whole division is observed in scattered manner and is mostly situated in the mid eastern part of the study area. This type of vegetation is found on the cool Landi plateau and moist localities near Mowar. The division is broadly divided into three physiographical formations:

   (i) High plateau with frequent undulating topography which approximately cover 60 per cent area.

   (ii) Moist sheltered valleys having deep fertile soil covering about 25 per cent area.

   (iii) Humid hills slopes, especially on northern aspect which covered approximately 15 per cent area of the division.

   This type of vegetation is mainly influenced by underlying rocks, which occur on soils derived from metamorphic rocks. These soils are heavier and compact in structure with greater moisture retentivity and are more suitable for miscellaneous species which have higher moisture requirements. Soils are usually dark brown in colour with some light coloured or reddish coloured patches. The red soils of the zone which have been formed in situ from indurate and metamorphosed rocks of Precambrian age on undulating to rolling topography. These soils are generally loam to sandy loam in texture, but may include gravelly in the uplands. They are generally natural but very slightly acidic to slightly alkaline in often reaction. Organic matter, N and P contents are low but Zn is high and clay content from 30 to 60 per cent. These soils are capable of growing large sized trees which are generally well formed and straight grown.

Plant species of this division has five main vegetation groups, ie 53 per cent trees, 21 per cent shrubs, 7 per cent climbers, 15 per cent herbs and 4 per cent grasses. The plant species as per Raunkiaer's Life-form classification are in existence as approximately 23 per cent *Mesophanerophytes*, 37 per cent *Microphanerophytes*, 13 per cent *Nanophanerophytes*, 7 per cent *Chamaephytes*, 3 per cent *Geophytes*, and 17 per cent *therophytes* (Table 3.6). Among post monsoonal trees species the leaf classes analysis show that they are 26 per cent *Macrophylls*, 33 per cent *Mesophylls*, 23 per cent *Microphylls*, 11 per cent *Nanophylls* and *Leptophylls*.

Comparatively, this division possesses the most favourable ecophysiographical conditions for the growth of plant species. Phytographically it is the richest division in the study area. As far as distribution of the trees is concerned Phytogeographically it is named as Anogeissus Terminalia type which is based on the frequency of their occurrence. Terminalia alata (Saja) and Anogeissus latifolia(Dhauda) are predominating trees. Other species are Lagerstroemia perviflora (Lendia), Diospyros melanoxylon (Tendu),Madhuca latifolia (Mahua),Terminalia belerica (Beheda),Salmalia malabarica(Semal), Soymida febrifuga (Rohan), Adina cordifolia (Haldu),and Pterocarpus massupium (Bija). These species form the top canopy while Dalbergial latifolia (Shisham), Grewia litiaefolia (Dhaman),Terminalia chebula (Harra), Bauhinia recemosa (Astera), Zizyphus zylophyrus (Ghatol), Cassia fistula (Amaltas), Emblica officinalis (Aonla) etc, are the middle canopy trees.

Table 3.6

Betul Plateau : Proportion of Life-Form in Various Vegetation Divisions

S.N	Vegetation Division	Mesopha-nerophytes	Micropha-nerophytes	Nanopha-merophytes	Chamae-phytes	Geo-phytes	Thero-phytes
1.	Tactona-Anogeissus Type	26	39	13	04	05	13
2.	Anogeissus-Terminalia Type	23	37	13	07	03	17
3.	Tactona-Lagerstroemia Type	14	36	21	06	05	18
4.	Chlorxylon-Diaspyros Type	13	41	17	07	03	19
5.	Boswellio-Sterculia Type	06	38	32	08	04	12
6.	Arjuna-Ficus Glomerata Type	11	36	17	13	04	19

Source: Based on Field Work.

Among shrubs Gymnema sylvostra (Gudmar), Vitexnegundo (Nirgundi), Lantana camara (Raimunia), Celastrus paniculata(Malkangni), Nytanthes arbortistis (Harsinger), Indigofera pulchella (Birhul) are dominants of this division. Abrus precalorius(Ghungchi), Bauhinia vahlii(Mahul), Combretum decandrum(Peeparbel), Tinospora cardifolia (Gilloy), Ventilago madraspatana(Kharbal) are common climbers.

Among sedge and grasses the dominants are Dutura metel (Dhatura), Sphaeranthus indicus (Gorakhmundi),Mimosa pudica (Lajwanti), Achyranthus aspera(Apamarg).Sitaria glavea(Kolia), Eulaliopsis binata(Sabai), Apludo mutica(Panai) and Cynodon dectylon (Dub), grasses are fairly distributed over the division. It is very interesting to mention here that Asparagus adscendens (Musli) a rare and valuable species is found at Bhawargarh hill. Musli herb generally restricted over the other divisions. This herb frequently observed over the humid slopes of compact soil area of Bhawargarh hills.

3. **Tectona-Lagerstroemia type** : It is the biggest vegetation division of study area which covers about 44 per cent area of Betul forests. It is obviously trap areas except the middle portion to south portion. The division is spread over the entire area of the plateau. Altitudinally two third of the division is situated up to 700 metres MSL. The whole division is divided into two main patterns of the physiographic formation: Regular and Irregular. The former formation is dominated over the Western, Southern trappean hills. The later formation is dominant over the undulating plateau of the study area.

   Soils of the division grouped under shallow and medium black soil category. These are light to medium textured and usually dark coloured (Haplusterts and vertic ustouchrepts) and reddish to yellowish brown. The soils of lowlying area are Typic Haplusterts, heavy textured and contain more than 40 per cent clay. These soils retain more water and nutrients. Upland soils are silty loam in texture and have water holding and nutrient retention capacity. On slopes the soil is richer and deeper.

The division has five main vegetation groups approximately 43 per cent trees, 7 per cent climbers, 17 per cent shrubs and 33 per cent herbs & grasses. The plant species as per Raunkiaer's life form classification are as approximately 14 per cent mesophanerophytes, 36 per cent microphanerophytes, 21 per cent nanophanerophytes, 6 per cent Chamaephytes, 5 per cent geophytes and 18 per cent therophytes.

More than 100 plant species are recorded from this division of which trees are the dominant species Tectona grandis (Teak) is the predominant species while Terminalia alata (Saj), Lagerstromia parviflora (Landia), Anoqaissus latfolia (Dhauda) are the codominant species. The main associates are, Diaspyros melanoxylon (Tendu), Emblica officinalis (Aonla), Boswellia serrata (Salai),Pterocarpus marsupium (Bija), Cassia fistula (Amaltas), Mitragyna parviflora (Kalamb), Bridelia retusa (Kasai), Aegle masmelos (Bel), and Bauhinia recemosa (Astra). (Table 3.7)

Shrubs commonly found are Woodfordia fruticosa (Dhabai), Helictres isora(Marodphal), Holarrhena antidysentrica (Dudhi), Indigofera pulchella (Birhul),Gardenia Gummifera (Dikamali), Abrus precatorius (Gumchi), Milletia auriculata (Gunja), Acasia pennata (Arail),Vallaris solanacea (Dudhbel), Smilax zeylanica (Ramdatun). Climbers are occasionally found.

Among herbs, Achyranthus aspera (Apomarg), Cassia tara (Chirotia), Hyptis suaveolens (Vantulsi), Tribulus terrcstris (Gokhar),Euphorbia hirta (Dudhi), Desmodium Latifolium (Lipti), are the dominant species. Among grasses the dominants are Heteropogon contortus (Sukal), Ischaemum laxum (Sainar), Lymbopogon martini (Rosha), and Cynodon dectylon(Dub).

Table 3.7

Betul Plateau : Profiles of Different Vegetation Divisions

S.No.	Vegetation Division	Physiographic/ Elevation	Location	Soils	Forest Physiognomy	Main Associates Plants Species
1.	Tactona-Anogeissus Type	Gently Sloping to undulating Plains up to 600 Metres	Machna, Tawa And Morand Sheltered Valleys, Lower Slopes of Bhanwargarh Hill.	Deep Black soils, clayey loam (Calcareous) Haplusterts	Miost Deciduous Forest	Tectona grandis, Anogeissus latifolia, Terminaliya tomantosa, Ougeniya oojeinensis.
2.	Anogeissus-Terminalia Type	Cold High Plateau Humid Slopes Low Lying Areas up to 700 Metres	Landi Plateau Near Mowar Paraskot, Matyardeo, Sirrikot Hills Slopes	Moderately Deep to Very Deep (Clayey) Haplusterts	Miost Deciduous Forest	Anogeissus latifolia, Terminaliya tomantosa, Diaspyros melanoxylon, Adina cordifolia.
3.	Tactona-Lagerstroemia Type	Undulating Plateau Trappean Hill up to 750 Metres	Chicholi - Multai Plateau, Bhanwargarh Kalibheet and Gawaligarh Hills	Shallow Medium Black (loamy) Medium Textured Black (Clayey) Ushochrepts	Dry Deciduous Forest	Tectona grandis, Lagerstroemia parviflora, Emblia officinalis, Cassia fistula.

4.	Chloroxylon-Diaspyros Type	Hilly Tracts, Rolling Topography 600 to 850 Metres	Asir Hill Khamala Plateau Tapti Valley	Light to Medium Textured, Red and Yellow Sandy Soil (Slightly Alkaline)	Dry Deciduous Forest	Chloroxylon swietenia, Diaspyras melanoxylon, Madhuca longifolia, Buchanania lanzan.
5.	Boswellia-Sterculia Type	Dry Tops of Hills, Rocky Slopes, Ridges. 500 to 900 metres	Southern escarpment Chicholi , Tapti Bhainsdehi Shallow Soil Areas Narthern Redges	Shallow Light Taxtured, Physiologically Dry Soils Ustochrepts	Dry Deciduous Low Forest	Boswellia serrata, Sterculia urens, Acacia leucophloea, Zizyphus jujuba, Acacia catechu.
6.	Arjuna-Ficus glomerata Type	Flood Plains Low Lying Lands up to 450 Metres	Banks of rivers And Nalas on Alluvial Patches	Alluvial, brownish (Clayey) Haplusterts	Potentially Evergreen	Terminalia arjuna, Ficus glomerata, Eugenia heyneana, Buta monosperma, Mangifera indica.

Source: Based on Field Work

4. **Chloroxylon-Diaspyros type** : Geographically it is termed as the division of hills and vegetationally by Hardwickia binata type which covers approximately 13 per cent area of the forests. It is spread throughout the sandy soil areas. Geographically it is spreading in northern part and southern part of the plateau. Altitudinally more than two third of the locality is situated 450 metres MSL. The division is broadly divided into two major physiographical formations:

   (i) Hilly sandy soils area with frequent undulating topography which approximately cover 70 per cent area.

   (ii) Scattered hillocks covering 30 per cent area.

   The soils of the division have been sandy, light to medium textured, slightly calcareous and slightly alkaline, in nature. The dominant soils of the division are Listochrepts, which are typified by the soils of Khirkiya series. These are highly deficient in nutrient with low storage capacity for moisture. The percentage of soil moisture has been increasing from pre to post monsoonal periods. Soil moisture regime is Typic Ustic and soil temperature regime is hyperthermic.

   The plants are distributed according to the nature of their physiographic formation. Many plants are observed either rare or abundant depending on the natural habitat. The plant species of the division is divided into five main vegetation groups; 48 per cent trees, 19 per cent shrubs, 8 per cent climbers, 18 per cent herbs and 7 per cent grasses. In post monsoonal period plant species of the leaf classes are 16 per cent *Leptophylls*, 17 per cent *Nanophylls*, 53 per cent and 14 per cent *Macrophylls*. The analysis of life form classification revealed that the region has 13 per cent plant species *Mesophanerophytes*, 41 per cent *Microphanerophytes*, 17 per cent *Nanophanerophytes*, 7 per cent *Chamaephytes*, 3 per cent *Geophytes* and 19 per cent *Therophytes*.

   The dominant vegetational associations among trees and shrubs Chloroxylon swietenia (Bhirra) is dominants tree. The other main associates being seen are Anoceissus latifolia (Dhauda), Diasphros melanoxylon(Tendu), Madhuca longifolia (Mahua), Lagerstroemia parviflora (Lendia), Emblica officinalis (Aonla), Albizzia odoratissims (Chichwa), Acacia leucophoea (Reunja), Lannea grandis (Moin), Gmelina arborea (Siwan), Stereospermum chelenoides (Padar), Terminalia balerica (Behera), Salmalia malabarica (Semal), Dalbergia paniculate (Phansi), Cassia fistula (Amaltas) and Kydia celycina(Baranga).

Nyctanthes arbortristis (Siharu), Indigofera pulchella (Birhul), Colebrookia appositifolia (Bhandar), Zizyphus zylophrus (Ghotal), Woodfordia fruticosa (Dhawai), Zizyphus jujube (Ber), Celastrus paniculatus (Malkangni), Phyllanthus niruri (Bhui Amla), Vitex negundo (Nirgundi), Holarrhera antidysenterica(Kutja) are common shrubs.

Among climbers,Bauhinia vahlii (Mahul), Combretum deeandrum (Peeparbel), Dioscorea hispida (Baichandi), Aegle marmelos (Bel), Smilex macrophylla (Ramdaton), Zizyphus rugesa (Churni), Helicteres isora (Marorphali), Vallaris solanacea (Dudhbel) are dominant.

Among sedges and grasses, Swerta anguslifolia (Chiretta), Achyranthus aspesa (Apamarg), Eranthemum purpurascens (Jungli tulsi), Echinops echinatus (Utakatara), Euphorbia hirta(Dudhi), Evolvulus alsinodes (Shankhpushpi), Boerhaavia diffusa (Punasnava), Acanthospermum hispidum (Gokharu) herbs are common while abundant grasses are Cymbopogon martini(Rosha), Sehima suleatum(Sainar) and Heteropogon contortus (Kusal).

5. **Boswellio-Sterculia type**: It is the smallest vegetation division of the Betul forests and covers only 10 per cent area. The whole division is observed in a scattered manner and mostly situated in southern portion of the study area. This vegetation division covers main three phytogeographical localities viz; southern escarpment, western hilly tract and tops of Sawaligrah hills under study region. Factors influencing the existence of this type vegetation in great abundance in these parts are the low rainfall, southern

aspect and shallow soil on rocky slopes. Altitudinally, the whole division is situated 450 to 750 metres from MSL. The division has three physiographic formations :

(i) On account of southern aspect

(ii) Ridges and

(iii) Dry tops of hills and rocky slopes.

Division consists of rocky habitat which is stony by nature and its colour varied according to the types of rocks. This type of vegetation colonizes sites on physiologically dry soils on all formation. These soils are shallow (5 to 20 cm deep) light textured, low in clay contents (about 10 per cent) and are dominated in illite and kaolinite type of clay minerals with low water holding capacity. These soils are poor in mineral matter, which are deficient in any fine particles, leading to soils with very little crumb structure and soils that reading break down, losing all cohesion.

It is concluded on the basis of field observation that reduced leaf size is one of the most important characters for the plant species of this division. Among post monsoonal plant species, the 'Leaf' classes are observed as 3 per cent leafless, 16 per cent *Leptophylls*, 31 per cent *Nanophylls*, 39 per cent *Microphylls* and 11 per cent *Mesophylls* and *Macrophylls*. The division has four main vegetational groups viz; 29 per cent trees, 7 per cent climbers, 23 per cent shrubs, 12 per cent grasses and 29 per cent herbs. The plant species as per Raunkiaer's life form classification are 6 per cent *Mesophanerophytes*, 38 per cent *Microphanerophytes*, 32 per cent *Nanophanerophytes*, 8 per cent *Chamaophytes*, 4 per cent *Geophytes* and 12 per cent *Therophytes*.

Boswellia serrata (Salai) is predominant tree of this division whose common associates are Anogeissus latifolia (Dhaura), Lannca grandis (Moin), Cochlospermum gossypium (Ganair), Chloroxylon swietenia (Bhirra), and Sterculia urens (Kulu). Occasionally teak is also found.

Among shrubs the dominant species are Zizyphus jujube (Ber), Aloe vera (Ghritkumari), Euphorbia nerifolia (Thuhar), Mucuna prurita (Kiwanch), and Opuntia elator (Nagphani). Sida cordifolia (Kharenti), Acacia cassia (Chilati), Cryptolepis buchanani (Karbel), Zizypus rugesa (Churni) are some of the common climbers found in this division.

There is always an abundance of grasses chiefly Heteropogon contortus (Kusal), Sehima nerovosum (Sainar) and Lymbopogon martina (Rosa). Herbs like Xanthium strumarium (Gokharu), Cassia tora(Chirota), Boerhavia diffusa (Punarnava) and Solanum suratience (Bhatkataiya) are common.

6. **Arjuna-Ficus glomerata type :** Geographically it is termed as the division of alluvial plains which covers approximately 6 per cent area of the Betul plateau. It has small patches and is scattered throughout the plateau. This type of vegetation extends right up to the water edges and on graval islands in the middle of the water courses. Altitudinally the whole division is situated from 450-500 metres from MSL. The division has two physiographic formations :

   (i) Flat and slightly undulating flood plains or banks of rivers which cover approximately 60 per cent area.

   (ii) Lowlying marshy areas which cover about 40 per cent area.

   The division is filled with recent alluvial materials, particularly near the banks of rivers. It consists of reddish to yellowish or brownish clay with numerous intercalations of sand and gravel. The constituents and the nature of the alluvium vary from place to place depending upon the rock composition in the catchment area from where it is derived. North of the hill is predominating in sand and sand material. In the soil of Tawa valley mantle is thin, poor and sandy except along the large water courses where it is deep and of finer texture. Sandy loam found along the banks of small streams and at the edges of Morand River. The area which receives washing from the Decan traps, the alluvium is predominantly clayey such as Tapti valley.

This forest forms a discontinuous canopy. The forest is not very dense. The various tree species are intricately mixed together. Individual tree may attain a height of up to 12 metres. It is usually open with tree growth. It is found on the flat terraces along rivers and streams, in the moist gulleys and on the flanks of river valleys.

The division has five main vegetational groups viz; 23 per cent trees, 19 per cent shrubs, 8 per cent climbers, 16 per cent grasses and 34 per cent herbs. It is concluded on the basis of field observations that most of the plant species in this division consist leaves with waxy surface. Among post monsoonal plant species the leaf classes are observed as 16 per cent *Nanophylls*, 27 per cent *Microphyll*, 25 per cent *Mesophyll*, 21 per cent *Merophylls* and 11 per cent *Macrophylls* or *Megaphylls*. The analysis of life-forms classification revealed that the region has 64 per cent plant species *Phaneorphytes*, 13 per cent *Chamaephytes*, 4 per cent *Geophytes* and 19 *Therophytes*.

Many plants are observed either rare or abundant depending on the nature of particular habitat but Terminalia arjuna(Koha) presents itself as an essential tree species in all the localities of this division. Among tree species Ficus glomerata Gular), Engenia heyneana (Katjamun), Butea monosperma (Palas), Mangifera indica (Aam), Tectona grandis (Teak), Dendrocalamus strictus(Bamboo) and Phoenix sylvestris(Chhind) are found as codominant species throughout the division.

Tamarix dioica (Jhau), Sida cordifolia (Kharent), Vitex negundo (Nirgundi), Lantana camara(Raimunia), Annona squamosa (Sitaphal), Ipomoea carnea (Sadabahar), Calotropis gigantean(Akauwa), Gymnosporia Montana (Baikal), Nyctanlhus arbortristis (Harsingar) shrubs are frequently found in this division. Climbers such as Abrus precatorius (Ghungchi), Butea superb (Palasbel), Tinospora cordifolia (Gilloy) and Asparagus recemosus(Satawari) are generally found.

Post monsoonal period is one of the most favourable seasons for the growth and development of ephemerals specially for the herbs and grasses. The dominant herb species are Cassia tora (Chirotia), Achysantles aspera (Apamarg), Argemone Mexicana (Pilikataria), Acanthospermum hispidum (Gokharu) ,Datura stramonium (Dhatura), Mimosa pudica (Lajwanti), Sphaeranthus indicus (Gorakhmundi), Chenopodium album (Bathua) and Jangli pudina.

Among grasses Sorghum helopense (Barru), Cenchrus ciliaris(Khus) and Cynodon dectylon (Dub) are common in the division.

ECOLOGICAL CLASSIFICATION

No where an organism or a species live alone. Always there are associates influencing each other and organizing themselves into communities. The organism of any community besides interacting among themselves, always have functional relationship with the external world or the environment. It is a functional system which in balanced condition is self sufficient and self regulated. In this system of plant community vegetation develops as a result of succession at one place of several seral communities which is completed when the succession of vegetation community often passing through different phases of changes culminates into equilibrium or mature condition. These phases of development of plant communities are characterized by maximum growth and development of plant. Thus, the plant community developed at the end of succession is called climax vegetation and is produced as a result of environment which includes climatic, edaphic, physiographic and biotic factors *(Tansley 1935, 291)*.

The climax is indicative of the stability of vegetation community of a given habitat, thus the climax denotes dynamic equilibrium of vegetation community which reveals equilibrium between not only vegetation community and its physical environment but also between all organisms *(Whittaker 1953, 151)*. A. Von. Humboldt has suggested that we are naturally led to consider each organism as a part of entire creation and to recognize it the plant or the animal, not merely an isolated species but a form linked to other form either living or extinct. Thus habitat of an organism represents a particular set of environmental conditions suitable for its successful growth. The development of different vegetation community takes place in different habitats and is affected and controlled by various factors. The distribution of plant type is directly related to the climate zone and both of them have reciprocal relationship. Occurrence of dominant species all over a climatic belt in various stages of growth and regeneration indicates that the area is under climax vegetation *(Clements, 1936, 256)*.

The plant communities modify the physical environment and provide the dominant influence. Since communities are distinguished by different life-forms and taxonomic composition of the dominant or predominant organisms, these charaterstics are usually used in naming the community. Since the community shows plant species variability and are not discrete entities that is why their classification has been a complicated problem. However, Godall (1965) has proposed a geometrical model to describe taxonomy of communities. The most widely used classification systems are based on physiognomy, habitat and floristic characterstics. The vegetation is sampled in small unit areas only and through these pictures of the entire community structure is formed in qualitative or precise quantitative terms. There may be many other ways in which dominant plants influence other members of the community. There is, therefore, an intimate interrelationship of all organisms in a community, some influencing profoundly others to a lesser extent. The mutual influences are both vertical and horizontal in communities. The physiognomic method basically divides the main plants present into various life forms irrespective of their taxonomic position. He further mentions that plant communities are said to posses a structure which reflects three components:

1. Stratification i.e. vertical distribution of different species occupy in different level,

2. The horizontal patterns of spatial distribution,
3. The abundance of each species.

The plant communities appear after the dominant form of life, that is why physiognomic classification is a highly useful method of lavelling and delineating the vegetation *(Hanson and Churchill 1961, 71)*. The different life forms influences other in a community and has its own ecological importance with respect to the rest of its members Life forms also indicates as to how a plant grows even in unfavourable conditions.

The ecological distribution of vegetation is designed with the opinion expressed by the above mentioned authors. Thus the following ecological classification is based upon the type of physiognomy. Very often the generic names of two or sometimes three conspicuous dominant are used to name each type of plant community. Therefore ecologically the vegetation of the area under study is divided into following classes: (Plate 3.4)

1. Teak forests
2. Mixed forests
3. Salai forests
4. Bamboo forests

GLOSSARY OF BOTANICAL AND LOCAL NAMES OF PLANTS FOUND IN THE BETUL PLATEAU

S.No.	Botanical Name	Local Name
1.	Acacia arabica	Babul, Kikar
2.	Acacia catechu	Khair
3.	Acacia pennata	Arail
4.	Acacia leucophloea	Hiwar, Ranjha
5.	Adina cordifolia	Haldu
6.	Aegle marmelos	Bel
7.	Ailanthus excelsa	Maharukh
8.	Albizzia procera	Safed Siris

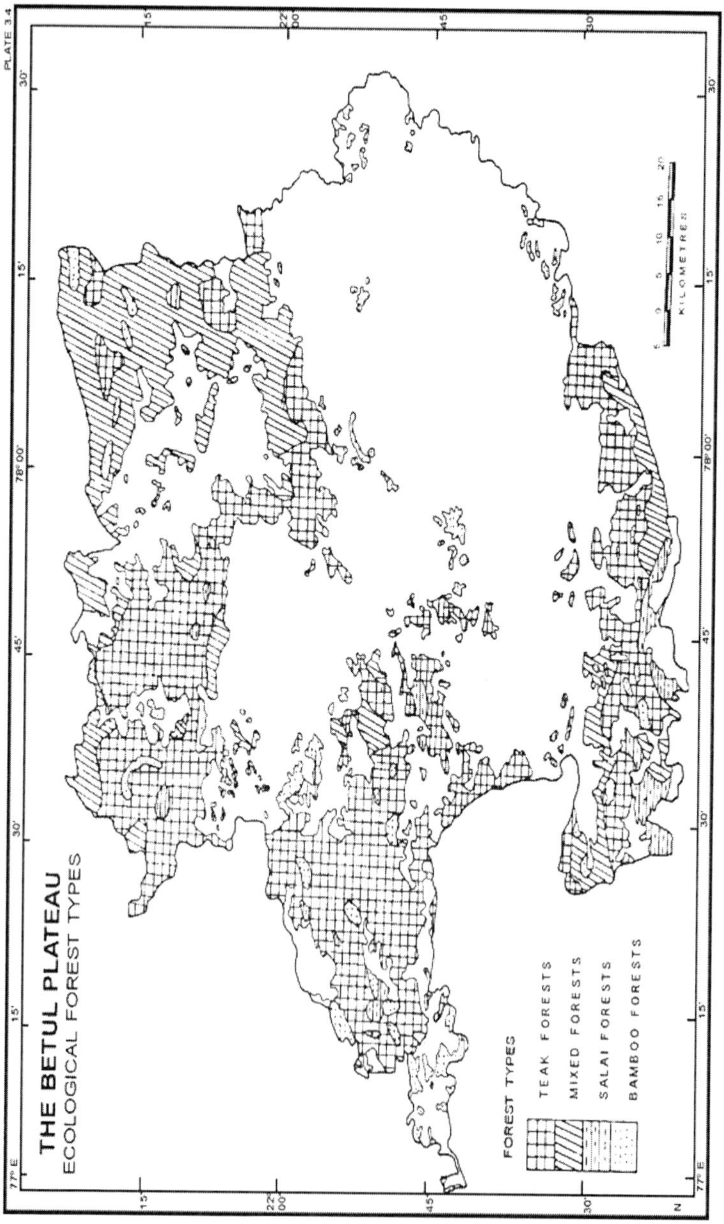
PLATE 3.4
THE BETUL PLATEAU
ECOLOGICAL FOREST TYPES
FOREST TYPES
TEAK FORESTS
MIXED FORESTS
SALAI FORESTS
BAMBOO FORESTS
KILOMETRES

9.	Albizzia labbeck	Siris
10.	Albizzia odorotissima	Chirol Black siris
11.	Anogeissus latifolia	Dhaora, Dharora
12.	Antidesma diandrum	Khatua, Katma
13.	Azadirachta indica	Neem
14.	Balanites aegyptiaca	Hingan, Hingote
15.	Bauhinia malabarica	Amta
16.	Bauhinia purpurea	Kealar
17.	Bauhinia racemosa	Astoo
18.	Bauhinia variegata	Kachnar
19.	Bosewellia serrata	Salai
20.	Bombax malabaricum	Semal
21.	Bridelia retusa	Kasai
22.	Buchanania lanzan	Achar
23.	Butea monosperma	Palas
24.	Careya arborea	Kumbhi
25.	Casearia graveolens	Gilchi
26.	Casearia tomentosa	Tondri
27.	Cassia fistula	Amaltas
28.	Chloroxylon swietenia	Bhirra
29.	Cochlospermum religiosum	Ganiar
30.	Cordia dichotoma	Lasoda
31.	Cordia macleodii	Dahiwas
32.	Dalbergia latifolia	Shisham
33.	Dalbergia paniculata	Dhobin
34.	Dilichandrone falcata	Medsing
35.	Diospyros melanoxylon	Tendu
36.	Ehretia laevis	Datrangi
37.	Elaeodendron glaucum	Jamrasi

38.	Emblica officinalis	Aonla
39.	Erythrina suberosa	Panjra
40.	Erythrina indica	Pangra
41.	Eugenia heyneana	Kathjamun
42.	Ficus bengalensis	Bargad
43.	Ficus glomerata	Gular
44.	Ficus infectoria	Pakar
45.	Ficus religiosa	Pipal
46.	Flacourtia indica	Kakai, Kanker
47.	Gardenia latifolia	Papra
48.	Gardenia turgida	Phetra (Safed)
49.	Garuga pinnata	Kekar
50.	Gmelina arborea	Siwan
51.	Grewia hirsuta	Gudsakari, Kareta
52.	Grewia tiliaefolia	Dhaman
53.	Hardwickia binata	Anjan
54.	Holoptelea integrifolia	Chirol
55.	Hymendictyon excelsum	Bhonrmal
56.	Ixora arborea	Lokhandi
57.	Kydia calycina	Baranga
58.	Lagerstromia parviflora	Lendia
59.	Lannea grandis	Moiyan, Gunja
60.	Madhuca latifolia	Mahua
61.	Mallotus philippinensis	Rohni, Sinduri
62.	Mangifera indica	Aam
63.	Miliusa tomentosa	Kari
64.	Mimusops hexandra	Khirni
65.	Mitragyna parviflora	Kalamb, Kalmi
66.	Ougeinia oojeinensis	Tinsa

67.	Phoenix sylvestris	Chhind
68.	Pongamia pinnata	Karanj
69.	Pterocarpus marsuprum	Bija
70.	Randia dumetorum	Mainhar, Mainphal
71.	Randia uliginosa	Phetra (Kala), Pindalu
72.	Samanca saman	Vilayti imli
73.	Salmalia malabarica	Semal
74.	Scheichera oleosa	Kusum
75.	Schrebera swietenioides	Mokha
76.	Semecarpus anacardium	Bhilwa
77.	Soymica febrifuga	Rohna, Rohun
78.	Sterculia urens	Kulu, Karrai
79.	Stereospermum personatum	Pader (Chhota)
80.	Stereospermum suaveolens	Pader (Bada), Paral
81.	Syzygium cumini	Jamun
82.	Tamarindus indica	Imli
83.	Tectona grandis	Sagon
84.	Terminalia arjuna	Koha, Arjun
85.	Terminalia bellerica	Bahera
86.	Terminalia chebula	Harra
87.	Terminalia tomentosa	Saj
88.	Terma orientalis	Jiwan, Thoar
89.	Wrightia tomentosa	Duddhi
90.	Zizyphus jujuba	Ber
91.	Zizyphus xylopyra	Ghatol
	SHRUBS	
1.	Aaave sisalane	Sisal
2.	Aloe vera	Ghritkumari
3.	Alangium salviifolium	Aukola

4.	Annona squamosa	Sitaphal, Sharifa
5.	Argemone mexicana	Siparkata, Shailkanta
6.	Calotropis gigantea	Akauwa, Madar
7.	Carissa spinarum	Karaunda
8.	Cassia auriculata	Tarwad, Tarwar
9.	Clerodendron viscosum	Bhandar
10.	Calanaregam spinosa	Menar
11.	Celastrus paniculatus	Malkangni
12.	Colebrookia oppositifolia	Bhandar
13.	Embelia ribesbrum	Baiberang, Baibedung
14.	Ensete superbum	Jangli kela
15.	Euphorbia neriifolia	Thuhar, Sehund
16.	Ficus hispida	Kathgular
17.	Gardenia gummifera	Dikamali
18.	Gymnema sylvestre	Gudmar
19.	Helicteres isora	Marorphali
20.	Hemidesmus indicus	Anantamul
21.	Holarrhena antidysenterica	Kutaja
22.	Indigofera pulchella	Birhul
23.	Indigofera tinctoria	Neel
24.	Jatropha camara	Ratanjot
25.	Lantana camara	Raimunia, Ghaneri
26.	Lawsonia inermis	Mehndi
27.	Leca macrophylla	Hathiakand
28.	Maytenus senegulensis	Bekal
29.	Mimosa rubicaulis	Chilati, Shiah-kanta
30.	Mucuna prurita	Kiwanch
31.	Murraya paniculata	Madhukamini
32.	Nyctanthes arbortristis	Harsingar

33.	Opuntia elator (dillenti)	Nagphani
34.	Phyllanthus niruri	Bhui Amla
35.	Plumbago zeylanica	Chitraka
36.	Sida cordifolia	Kharenti
37.	Tamarix dioica	Jhau
38.	Vitex negundo	Nirgundi
39.	Woodfordia fruiticosa	Dhawai
40.	Zizyphus oenoplia	Makai
	HERBS	
1.	Achyranthus aspera	Apamarg
2.	Acanthospermum hispidum	Gokharu (Chhota)
3.	Argemone mexicana	Pili katari
4.	Asparagus adscendens	Musli (Safed)
5.	Boerhaavia diffusa	Punarnava
6.	Cassia tora	Chirotia, Panwar
7.	Chenopodium album	Bathua
8.	Datura metel	Dhatura (Kala)
9.	Datura stramonium	Dhatura (safed)
10.	Desmodium latifolium	Lipti
11.	Echinops echinatus	Utakatara (Utakanta)
12.	Eclipta alba	Bhringraj
13.	Eranthemum purpurascens	Jungli Tulsi
14.	Euphorbia hirta	Dudhi
15.	Evolvulus alsinoides	Shankhpushpi
16.	Hyptis suaveolens	Vanmtulsi
17.	Mentha arvensis	Podina
18.	Mimosa pudica	Lajwanti
19.	Ocimum sanctum	Tulsi
20.	Sesamum orientale	Til

21.	Solanum suratiense	Bhatkatiya
22.	Sphaeranthus indicus	Gorakhmundi
23.	Swertia angustifolia	Chiretta
24.	Tribulus terrestris	Gokhar (Chhota)
25.	Xanthium strumarium	Gokhru
	CLIMBERS	
1.	Abrus precatorius	Ghungchi
2.	Acacia Cassia	Chilati
3.	Acacia pennata	Arail
4.	Acacia rugata	Shikakai
5.	Acacia torta	Arial
6.	Asparagus recemosus	Satawar
7.	Bauhinia vahlii	Mahul
8.	Butea superba	Palas bel
9.	Cissus quadrangularis	Hadjora
10.	Clematis triloba	Murhari, Morbel
11.	Combretum decandrum	Peeparbel
12.	Cryptolepis buchanani	Nagbel
13.	Dioscorea bulbifera	Ratalu
14.	Dioscorea hispida	Baichandi, Karukandu
15.	Ichnocarpus frutescens	Kalidudhi, Dhimarbel
16.	Jasminum arborescens	Chameli
17.	Millettia auriculata	Gurar
18.	Momordica dioica	Kakadbel
19.	Mucuna prurita	Kiwanch
20.	Smilax zeylanica	Ramdatun
21.	Tinospora cordifolia	Gilloy
22.	Vallaris solanacea	Dudhbel
23.	Ventilago calyculata	Kevati

24.	Ventilago madraspatana	Kharbel
25.	Zizyphus rugosa	Churna, Suran
	GRASSES & BAMBOOS	
1.	Apluda mutica	Ponai
2.	Arundinella selosa	Phoolbahari
3.	Bothriochloa pertusa	Mekhal
4.	Cenchrus ciliaris	Anjan, Kus
5.	Cynodon dectylon	Dub
6.	Cymbopogon martinii	Rusa
7.	Dendrocalamus strictus	Bans
8.	Dichanthium annulatum	Marbel
9.	Eragrostis tenella	Bhurbhusi
10.	Eulaliopsis binata	Sabai
11.	Heteropogon contrortus	Sukal
12.	Iscilema laxum	Muchbel
13.	Pennisetum pedicellatum	Dinanath
14.	Saccharum spontaneum	Kans
15.	Sehima sulcatum	Sainer
16.	Sitaria glauca	Kolia
17.	Themeda quadrivalvis	Gunher
18.	Vetiveria zizanioides	Khas
	EPIPHYTES & PARASITES	
1.	Cuscuta reflexa	Amarbel
2.	Dendrophthoe falcata	Banda
3.	Vanda parviflora	Orchid
4.	Viscum articulatum	Banda

CONCLUSION

The plateau, surrounded by the mountain ranges at its all the three sides has a special physiographic personality which dominants the forests ecology. The area is located within the tropical belt with a monsoon type of climate. Due to topographic variation, sometimes micro-climatic changes occur in the area. The Satpura range of the region towards east-west in the northern part receives maximum rainfall. On the other hand rain shadow areas are Bhansdehi-Multai plateau to the south of the Mahadev hills, northern part of the Gwaligarh hills to the south so these areas are dry. The physiographic features have made an easy ascent of Arabian monsoon in the northern part with the consequent wetter conditions compared to the southern part. This rainfall pattern is in turn manifest in the distribution of the vegetative cover.

1. More than 100 plant species of Betul plateau under natural vegetation is recorded and analysed for their vegetational characterstics but an overall general account is given here for the plant species spectrum as a whole. The plant species of the region is divided into six main vegetation groups i.e. 43 per cent trees, 19.3 per cent shrubs, 7.3 per cent climbers, 20.8 per cent herbs and 9.3 per cent grasses. The analysis of life forms classification revealed that the region has 15.3 per cent plant species *Mesophanerophytes* (Over 10 metres tall), 38 per cent *Microphanerophytes* (2-10 metres), 18 per cent *Nanophanerophytes* (0.5-2 metres), 7.4 per cent *Chamaephytes* (Below 0.5 metres), 4.5 per cent *Geophytes* and 16.3 per cent *Therophytes*. In post monsoonal period among plant species the leaf classes are as 14 per cent *Microphylls* (18225-164025 sq. mm.), 17 per cent *Mesophylls* (2025-18225 sq. mm.), 42 per cent *Microphylls* (225-2025 sq.mm.), 16 per cent *Nanophylls* (25-225 sq. mm.), 8 per cent *Leptophylls*. Thus Life forms spectrum of the plant species for this region shows that the plant species are *Phanerophytic* in nature.

   Vegetation of the region is of deciduous nature. The north-east part has subtropical wet hill forests while tropical dry deciduous forests grow towards north-west, west and south portion. The high hill tops of the Mahadev hills are characterized by the moist high level forests while lower humid slopes and valleys have moist low level forests.

2. Geology of the area also plays an important role in the distribution and composition of the plant community. The different types of forest mainly thrive in the Deccan trap and Gondwana formation. The southern and western part of the region with Deccan trap formation has extensive tropical dry deciduous forests whereas north-east and central part with rocks of Gondwana system, metamorphic, is characterized by moist low and high level forests respectively. The trap supports teak species while in sandstone areas miscellaneous species grow well.

3. The soil is another factor which affects the type and distribution of plants directly in the study area. The composition of flora varies with soil variation. The depth of the soil changes according to the changes in the slopes. The trap soil is found in wide plain areas which are covered with dry deciduous forests. The forest areas with good and fertile black soil consist of good quality crop. The alluvium of river are rich in moisture content, favours good quality of moist deciduous forests. The north-east part with thin mantle of soil has poor quality forests. In this way soil and vegetation relationship is well marked.

   The study area has about 40 per cent forest area of which nearly 87 per cent forests area is extended over hills, high plateau and undulating areas. In northern part, near about 50 per cent area falls under forests and remaining proportion of the forests spread over the western and southern hills and high plateau areas. The extensive forest stands are found in the ravines of the Tapti valley to the central part of the study area and on undulating land parts. The trap covered mid-eastern and south-eastern flat plateau area is used for agriculture. The river valleys of the region are densely forested with good quality of crop. The top of high hills and lower slopes are covered with moist forests. The northern and southern escarpments and ridges have poor quality forests.

REFERENCES

Botkin, D.B. and Kellar, E.A, 1982, *Environmental Studies,* C.F. Merrill Publishing Company, Columbus, A Bell and Howell Company.

Champion, H.G 1936, A Preliminary Survey of forest Types of India and Burma, *Indian Forests Research Journal,* (NS)(Silv.), Vol. I, (i), 1-286.

Champion, H.G. and Seth, S.K, 1968, *A Revised survey of the forest types of India,* New Delhi, Manager of Publications.

Clements, F.E. 1936, Nature and Structure of the Climax, *Journal of Ecology,* Vol. 24, 252-284.

Davis, P., 1979, *Science in Geography, data Description and Presentation,* London, Oxford University Press, Vol. 3, 60-61.

Goodall, D.W., 1965, Objective Methods for the Classfication of Vegetation, *Journal of Botany,* Vol. I(8), 39-62.

Hanson, H.C. and Churchill, E.D., 1965, *The Plant Community,* New York, Reinhold Publishing Corporation.

Humboldt, A.Von., 1807, Ideen Zu einer geographic der pflanzen nebert einem Naturgemalde der Tropeolander, Tubingen.

Raunkiaer, C., 1934, *The Life Form of Plant and Statistical Plant Geography,* being the Collected Paper of C. Raunkiaer Oxford, Clarendon Press.

Tansley, A.G., 1935, The Use and abuse of Vegetational Concepts and terms, *Ecology,* Vol. 16, 284-307.

Tivy, J., 1982, *Biogeography,* Longrnan, London.

Whittaker, R.H., 1953, A Consideration of Climax theory-the Climax as a Population and Pattern, *Ecological Monographs,* Vol. 23, 41-78.

4
Ecological Status of Teak Forests

The Betul plateau part of Satpura is a delicate high land eco system. This ecological fragility is rooted in its geological and morphological setting. This ecosystem evolved its own self-defense over millions of years in the evolution of the dense natural forests that sheltered the slopes from the direct impact of the torrential rainfall and facilitate the absorption of water for delayed discharge via springs and streams. This central and very significant ecological process is gifted with a very wide variety of biotic resources rich in animal and plant heritage. About 40 per cent land of the study area is covered with forests. The forest cover is a sensitive indicator of regional environmental quality and well being. Forests can enrich human life in variety of ways both material and psychological.

In the study area economically most important teak grows by dint nature which is predominant and prevailing specials in the most valuable teak forests. Betul teak is famous all over the country and compares well with the Burma teak. Teak forest of this region is very rich and produces a variety of products. Differents types of timber are major produces of this forest in the study area. Because of the variation in composition of forests a variety of wood is obtained, however teak constitute the bulk of the output. After teak, Terminalia tomentosa (Saj), Pterocarpus marssupium (Bijasal), Adina cordifolia (Haldu), Daibergio latifolia (Sisham), Ougeinia oojeinenensis (Tinsa) are other important species, used for timber. Bamboo abounds in teak area. Amongst the main forest products of the study area bamboo occupies second place. Teak, other timber wood and bamboo for which the demand is quite keen yield bulk of the forest revenue. They are excellent quality largely exported and used for building and industrial purpose.

The Betul forests are not only important for their economic utility but they also influence the social and economic life of mankind. The benefits occurring to inhabitants from forest are various and derived in a variety of ways. Agriculture, animal husbandry and its allied activities are also dependent on forests to a large extent. The agricultural implements are made from wooden poles and bamboos. Bamboo is used for fencing, basket making, and tool handles. Thus, furniture has great potential in boosting the economy of primitive tribale of study area.

Ecologically teak forests area is grouped into moist and dry type deciduous nature *(Champion, 1936, 1-226)*. This classification is based mainly on climatic and ground conditions. The moist teak forest occurs in area having rainfall above 50 inch (127cm.) while medium or deficient rainfall is responsible for the occurrence of dry type of forests. In these forests teak (Tectona grandis) is the principal species among the high trees, because of its extraordinary surviving capacity. It is found as pure crop as well as in association with other species in varied proportion. Its physiographical distribution is being influenced mainly by the nature of the underlying rock form which the soil is derived, eco-climate and topography.

ECOLOGICAL REQUIREMENTS

Eco Climate

Although teak forest occurs in dry localities subject to high temperature and drought in the hot season, it thrives best and reaches its largest dimensions in fairly moist, warm tropical climate. As for temperature in the teak zones of Indian Peninsula where pure teak forests stand, the maximum and minimum shade temperature are 48°C to 3°C. In the study area annual maximum and minimum temperature are 42.7°C and 5.6°C with 23.8°C as annual mean. The highest maximum temperature and lowest minimum temperature never exceeds the critical limit and is therefore not limiting factor.

Peninsula teak forest appears to thrive best in regions having a normal rainfall varying from 75 cm to 400cm. In the study area normal annual rainfall ranges from 76.8 to 158.2cm. so as to avail the full benefits for good growth of teak forest. Eco-climate of Betul Plateau varies month to month which are as follows (Table 4.1):

Table 4.1

Betul Plateau : Eco Climate Classes

Temperature & Moisture Class	Eco Climate Type	Month
1. Temperature Class	Hot	April, May, June, September
	Warm	Jan., Feb., March, July, Aug., Oct., Nov. and Dec.
2. Atmospheric Moisture Class	Desiccating	Jan., Feb., March, Nov., Dec., April, May
	Dry	-
	Humid	June, Oct.
	Damp	July, Aug., Sept.

It is concluded on the basis of field observations that the best months for vegetation growth of teak forests are June to October (five months) and poor months for vegetation growth are November, December and January to May (seven Months). Thus the annual rhythm of its growth fluctuations coincides to the prevailing climate conditions.

Geology and Soil

Teak forest is found on all geological formation, except the Gondwanas but it is intimately associated with and thrives exclusively on the Deccan Trap and the resultant soil. Soil produced by trap rock being fairly sandy loam on the slopes and deep clayey loam on the low lying lands with sufficient minerals to supports a fairly good forest crop. The proportion of teak on traps has increased to almost pure forms.

The crystallines formation in contrast has a lesser percentage of teak which attains large size in the valleys, where the soil is deep, fertile, well drained loam or clayey loam. In part of Gawasen range teak may be found as pure crop. Soils derived from metamorphic and conglomerate rocks teak forests are generally poor in growth as well as proportion (5 to 30%) with other species of mixed forest. Teak forest of superior quality occurs on the narrow alluvial strips along the Tawa, Morand and Tapti rivers.

Geomorphology

The majority of teak forests are situated on hilly or undulating terrains, but there are considerable teak areas on well-drained flat alluvial grounds. On well drained alluvial plains teak sometimes occurs in its remarkable pure form and attains large dimensions. It also attains excellent development on the lower slopes of hills but along dry ridges it becomes stunted.

DISTRIBUTION OF TEAK FOREST

Economically important teak forest spread over 1853.9 square kilometres. They cover 18.4 per cent of total geographical area and 46 per cent of the total forest area of the area under study. It is concentrated in north division with 33.03 per cent of the forest area of the study region which is about 50 per cent of the forest division. The south division has 35.48 per cent of the total forest area and 41.8 per cent of the forest division. Distribution of area under teak forests in the different divisions is given in Table 4.2.

In north part of the study area teak forests spread over the southern slope of the Satpura range. It is found on Bhanwargarh hills of Barbatpur range, the whole part of Bhaura range leaving northern spur and in the forest area of Gawasen and Betul range.

Table 4.2

Betul Plateau : Distribution of Teak Forests

Forest Division	Total area under forests (in hect.)	Total area under teak forests	Teak forests as% of total area (in hect.)	Percentage of regional area under teak forests
North Division	1,23,215.80	61,251.40	49.71	33.08
South Division	1,57,078.00	65,785.69	41.88	35.48
West Division	93,112.00	56,690.00	60.88	30.57
Rampur Bhatodi Project	30,729.50	1,667.23	05.42	01.00

Source: Compiled from records of the office of the conservation of forests, Betul 1998.

The southern portion consists of teak forest in the Tapti valley, in northern part of Gawaligarh hills and high plateau area along these hills. In western portion teak is found in the Tapti valley, on Saoligarh hills and Tapasari peak of Kalibheet hills.

Teak forests are found in all the 15 ranges and Rampur Bhatodi Project area of the study region. The teak growing ranges have been arranged into 5 categories following the method of Davis (1979, 60-61). The categories are high, medium high, medium, medium low and low. High percentage of teak forests is found in Gawasen, Tawdi, Betul, Athnair, Saoligarh, Chicholi, Mohda, Barbatpur,Bhaur and Tapti, Whereas low percentage is found in Rampur Bhatodi Project (Plate 4.1). Distribution in other ranges is presented in Table 4.3

The quality, composition and the condition of the crop growth varies considerably from one place to the other and a few important ranges are therefore described separately.

The Barbatpur Range

In the northern part of the region forests of Barbatpur are known for good quality teak. This range spreads over Bhanwargarh hills in the north of the study area to the west of the Tawa River. It consists of about 55 per cent teak forest land of total forest area. The Machna valley in the middle portion of the range has moist teak forests while Chilpokna, Bijadehi and Southern Bhanwargarh hills of the southern Satpura range are covered with dry type forest due to occurrence of sandy soil with sand stones. On the lower slopes and in the valleys the percentage of teak stands as 80 per cent of the growing crop. Pure teak stands may be seen as in the case of small island in the Machna.

The range is extended over 4026.39 kilometres and consists of 38 per cent area under teak of total area. The depth of alluvium varies at different places and so causes variation in the density of crop. In general, the density ranges between 0.5 to 0.6 of dry type and 0.9 of moist type teak forests.

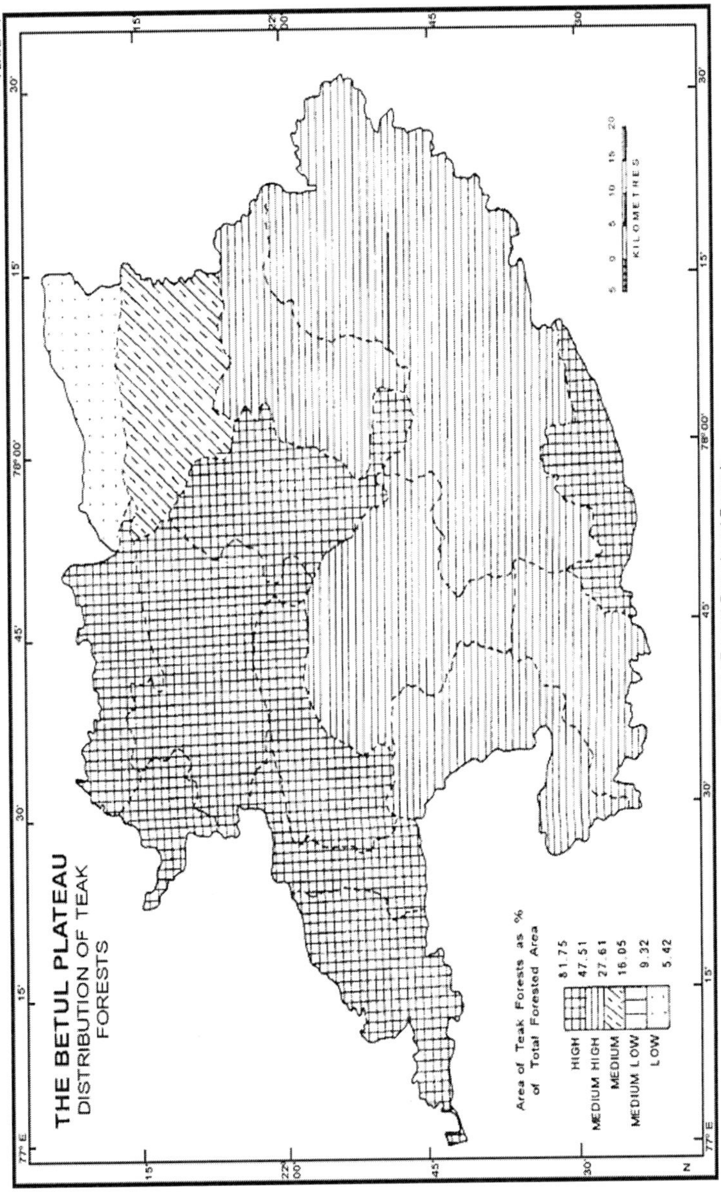

Source : Forest Department Records

The Betul Range

The middle portion of the plateau comes under this range which is scattered over residual hills, of Satpura range to the north and over the trap covered high plateau area to the south. The area of the ranges is 21659.7 hectares. Out of this area more than 60 per cent is under teak forests. The teak grows 50 per cent of growing crop. This part of study area receives heavy rainfall (1195mm.) so causes occurrence of good quality moist teak forests. In some parts of the range such as north-east part with Gondwana system rocks, exposed slopes, flat hill tops, on higher elevation and on spur, the moist type teak forests degenerate into dry type.

The Gawasen and Saoligarh Range

The range is confined to the north-east part of the study area in 182 square kilometres and consist 82 per cent land of teak. In this range, the northern land of Saoligarh, Chauradeo region, Harrai, Gawasen and Jugmadeo hills are densely forested with teak. The higher elevation consists of dry type teak forests due to the presence of red gravel and thickly strewn with reddish stones mixed soil of less humidity. The extensive teak areas are the river valleys of Bhangi, Dudhi and Larbaria. The quality of crop is good with well formed cylindrical trees. The teak is found 50 per cent on an average and reaches above 50 per cent on lower slopes. The density varies from 0.5 to 0.9 to almost full.

The Saoligarh range spreads over 25 per cent land of the study area to the west. The percentage of teak is above 80 of the growing crop. The Saoligarh hills in the south and the Narwargarh hills to the north of the range are the extensive teak areas. The teak is mainly confined to the Dhodhrama, Phoplia and Khokhara kheda hills. The quality of the crop is good in the ravines of the river valleys of the Morand and its tributaries, in Tangna, Pandijhiri and south Phoplia forest areas.

The Bhaura Range

The range is extended over the north Satpura range to the north-east part of the plateau in 306.4 sq kilometres of land. The teak covers 50 per cent land of the range (Plate 4.2). In the southern part of the range the crop grows as a belt from west to east. The southern slopes of the Satpura range are eroded by Morand and its tributaries so covered by extensive teak forests. The Bhanwargarh hills to the east of the Satpura range, Chikwarbhilra Pahar and spur of residual peaks scarping to the river valleys are densely forested with teak. In the western part of the range the teak spreads in a wide area of Pakshi and Pat forest areas. The quality of teak is good in valleys of river Morand, Bhangi and their tributaries. The good quality of teak is mainly confined to the northern slope of Satpura range covered with trap because of heavy rainfall. The percentage of teak is above 50 of growing crop. The density is 0.6 to 0.7 in general but it varies up to 0.8 in Bhanwargarh range and ranges of western boundaries.

The Tapti Valley Ranges

The teak forests are scattered over undulated Tapti valley and high plateau area along the valley in the mid-western part of the study area. The range is extended over 962.9 sq kilometres to the south of the Chicholi plateau along both the sides of the Tapti valley and consists of 47 per cent teak forests. The crop may also be seen in the valleys of the Betul and Nishan River to the north of the river Tapti. The deep alluvium of the valley area provided with moist type of teak forests but land area consists of dry type teak forests. The good teak forests can be seen in the alluvium of regional rivers and in Amdhana & Chunlama. The density usually varies from 0.5 to 0.6.

In Tapti valley, the Tawadi range extends over 368 sq. kilometres to the north of high land of the Chicholi plateau. Here, teak forests are scattered as small blocks in the ravines of Tapti, Batki, Labada and Khandu and its tributaries to the north and south respectively. To the north of the range Godhakhar hillocks and to the south of Takjhiri and Dodajam, the teak areas are extensive with poor quality of crop as these areas are greatly eroded and consist sandy soil. The density ranges between 0.3 to 0.4.

The Mohda range spreads over 607 sq. kilometres along both

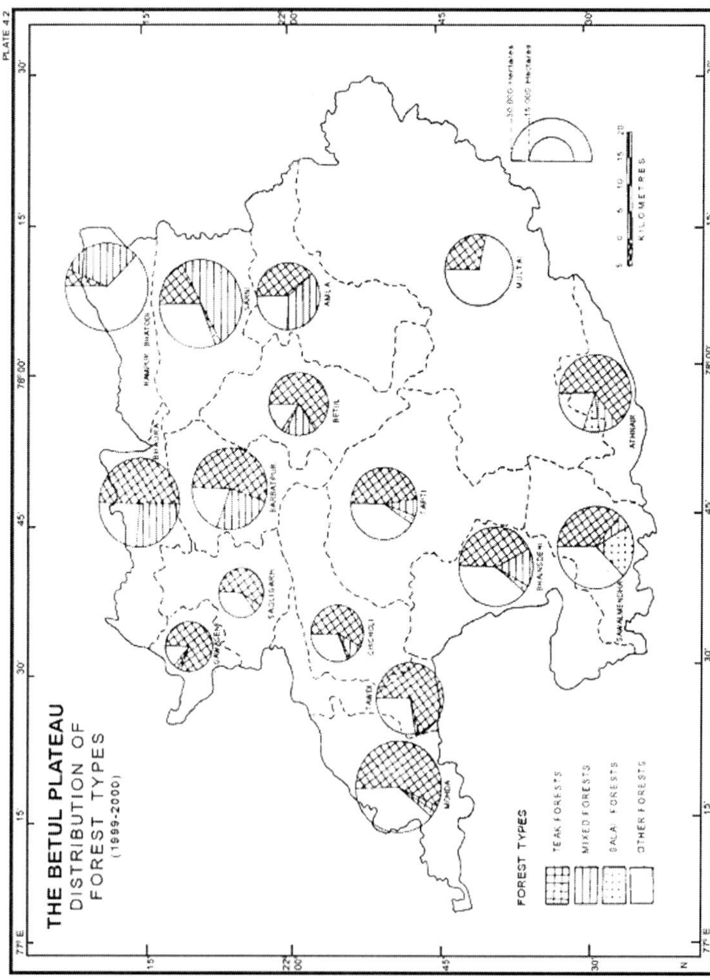

Source : Foreest Department Records

the sides of the Tapti valley from river Batuki to the western boundary of the plateau. The teak covers 30 per cent land of the total forest area and the density is 0.4 to 0.5. The crop is mainly confined to the Babrideo hills and Tapti valley to the north and south respectively.

The Ranges of Gawaligarh Hills

The Bhainsdehi, Sawalmendha and Athnair ranges of south division are included in the ranges of Gawaligarh hills. These ranges are extended from the southern part of the river Tapti to the Gawaligarh hills. The teak forests may be seen in Patoli to the south on Gawaligarh hills and in the northern part on Neelgarh hills and in the Tapti valley in Sawalmendha range, the Tapti valley to the north, high plateau area, hills to the south and upstreaks of Purna and its tributaries are teak forest areas. In Matka, Tanki and Sawalmendha forest areas the percentage of teak is 30 to 50 of growing crop.

In Athnair range the teak forests cover 60 per cent land part (Plate 4.3). In this area the teak forests of Dabka are scattered over Hirli, Salbardi and Saibardi hills. The percentage of crop is more than 50 of growing crop. The density varies mainly from 0.5 to 0.6.

Other Ranges

In the Sarni range; the Tawa River forms a deep valley which is protected from wind and temperature near Mahadev hills to the north and high plateau area to the south. The humid slopes of this range consist of sand mixed loamy soil from Talchir and Barakar rocks of Gondwana system. This area is occupied by moist deciduous teak forests. These forests also occur on marginal part of the Sarni Range to the east where depositions of trap soil are found.

The poor quality crop grows on Tapasari hills of eastern Kalibheet hills and in the region of brown triabic loam in south-west Chicholi plateau.

STRATIFICATION OF THE TEAK FORESTS

Stratification of the land communities reaches the greatest

Betul Plateau : Teak Forests

Betul Plateau : Bamboo forests

complexity in the forest community. In forests, different species of plants are evolved in the habitat having favourable environmental conditions. This takes place through the process of adaptation, competition and natural selection. This resulted in the development of various strata or layers. The formation of these layers is called stratification in which different species are found at different levels.

Raunkiaer (1934, 639) has classified plants on the basis of their life forms which is the result of relationship between plant life and climatic factors. He considered stratification as an index of life forms. On the basis of stratification, the life form has four layers. Firstly, dominant layer determined by the canopy of the largest trees. Second layer is located below first one and is represented by shrub by life forms, third layer is called herb and fourth layer represents mosses on the ground thus called strata or ground layer.

The teak trees form the upper storey of the teak forests. The variation may be seen in the height of trees from place to place according to the depth of soil and slopes of the land.

The main associates of teak in the top canopy includes Terminalia tomentosa (Saj), Adina cordifolia (Haldu), Lagerstromia parviflora (Lendia), Ougeinia oojeinensis (Tinsa), Madhuca latifolia (Mahua), Mitragyna parviflora (Kalamb), Miliusa tomentosa (Kari) and Syzygium cumini (Jamun) that are evident in the Tawa valley. Lagerstromia parviflora (Lendia) and Ougeinia oojeinensis (Tinsa) are common in Machna valley. Bosewellia serrata (Salai) is found only on higher elevation. Some species viz; Grewia tiliaefolia (Dhaman), Madhuca latifolia (Maiyan),Terminalia bellerica (Baheda), Schrebera swietenioides (Mokha), Scheichera oleosa (Kusum),Cassia fistula(Amaltas),Ficus glomerata(Gular), Bombax malabaricum(Semal),Aegle marmelos (Bel), Diospyros melanoxylon (Tendu),Dalbergia latifolia (Sheesham), Pterocarpus marsuprum (Bija) are found less frequently. Middle storey is often composed of Emblica officinalis (Aonla), Flacourtia indica (Kekai),Buchanania lanzan(Achar), Soymica febrifuga(Bhelwan) etc. Sterculia urens (Kuku) is found along sheltered valley and along nalas in low lying areas of felling series. In dry type teak, forests are generally single storeyed and where it is not, trees found in the understorey are very few. Bamboo is found along the slopes and is absent on the hill tops and the flat plateaus.

Common shrubs found in the undergrowth are Nyctanthes arbortristis (Siharu), Helicteres isora (Marodphali), Woodfordia fruiticosa (Dhabai), Indigofera tinctoria (Neel), Sida cordifolia (Khareta), Carissa spinarum (Karaunda), Argemone Mexicana (Siparkata), Aloe vera (Ghritkumari), etc. Lantana camara (Raimunia) has invaded in several areas forming a thick cover and is spreading fast.

Herbs occurring occasionally are Cassia tora (Chirota),Xanthium strumarium(Gokharu),Hyptis suaveolens(Van Tulsi), Desmodium latifolium (Lipti), while grasses like Heteropogon contrortus (Sukal), Themeda quadrivalvis (Gunari)and Sehima sulcatum(Sainar) are commonly found. The existing undergrowth varies with the density of the canopy. Climbers are generally absent, only species like Acacia pennata(Arail),Acacia Cassia(Chilati), Abrus precatorius (Gumchi) Bauhinia vahlii (Mahul) and Butea superba(Palas Bel) are occasionally found. Teak being an exacting species does not allow much undergrowth excepting bamboo to come under it.

VALUABLE SPECIES

Tectona grandis Linn/Teak Tree

The teak is the 'royal tree' of India. It is an important timber tree which grows in natural form in deciduous forest belt of the country.

A Large deciduous tree with a rounded crown and under favourable conditions a tall clean cylindrical bole. Its bark is grey or light greysish, brown fibers with shallow longitudinal fissures. Sapwood small, whitish, heartwood dark golden yellow, sometimes with dark streaks, turning brown with age. Branchlets quadrangular, channeled, with large quadrangular pith. The leaves are about 30 to 50 cm long, broadly elliptical or obviate in shape. The old leaves sheed in winter and the tree remains leafless till the new leaves appear in summer. Flowers are bluish white, flowering season is rainy and the fruits ripen in winter. From Life-forms point of view it comes in the group of *Megaphenerophytes* and leaf class of the tree again fall into class of Mecrophylls.

Natural Habitat

Although teak occurs in dry localities subject to high temperature and drought in the hot season. It thrives best and reaches its largest dimensions in a fairly moist, warm tropical climate. Teak grows in places where temperature ranges between 5°C to 48°C. It appears to thrive best with a normal rainfall varying from 75 cm to 400 cm. The teak requires good subsoil drainage. On black cotton soil teak occurs remarkably pure and attains large dimensions but very sandy or gravelly soils are unfavourable site for the growth of the teak. Teak crop is also rich in alluvial ground & on the crystalline areas. Thus, by nature we can say that the tree shows poly climax tendency of plant succession.

Phytogeoraphical Distribution

This tree is widely distributed over peninsular of Indian sub continent including nearly all types of habitats. Although the teak reaches large dimensions in some of the forests of western and southern India. The teak belt runs northwards upto Barauch district in Gujarat along the Sahyadri range running along the coastal parts of Kerala, Karnataka and Maharashtra. The belt then turns from Thana eastwards upto Balaghat and Jabalpur district of Madhya Pradesh forming a broader zone of more than 150 kilometeres across Nasik, Dhulla, Kargone, Jalgaon, Khandwa, Buldhana, Hoshangabad, Betul, Amravati, Chhindwara, Seoni and Balaghat districts. The teak belt then bifurcates into two branches- one extending northwards upto the Jhansi district of Uttar Pradesh and the other running southwards upto east and west Godavari districts of Andhra Pradesh. The southern off shoot the zone is relatively a narrower belt while the northern offshoot is broader. There is another isolated patch of teak forests in the Karnool district of Andhra Pradesh.

The great majority of the teak in this belt is occurred on two great rock systems (i) The Decan trap which stretches in the southern

Indian states teak forests are found in the western parts of Karnataka and Andhra Pradesh and (ii) in the areas of crystalline rocks (granite,gneiss,schist)in Bundelkhand,several parts of central India, the western parts of Andhra Pradesh, the Western Ghats and throughout the greater part of Tamil Nadu. On the Decan traps, where the soil often very superficial, the teak is usally of small size but occurs in great abundance, often forming the bulk of the growing stock and great abundance, often forming the bulk of the growing stock and even occurring pure over considerable areas.The crystalline areas, although the teak trees are as a rule more scattered that they are on the traps, they reach considerably larger dimensions where the rainfall is favourable.

In the Jhansi area of Uttar Pradesh, teak occurs on geneiss and quartzite and is confined to areas within a few miles of the larger rivers. In Madhya Pradesh, teak occurs to a greater or lesser extent in Jabalpur, Damoh, Sagar, Hoshangabad, Seoni, Chhindwara, Balaghat, Raipur and Betul districts, In Maharashtra, it occurs in Nagpur, Wardha, Chanda, Buldana, Amravati districts and possibly to a small extent in other localities of peninsular India. In Balaghat and Raipur, it is very local, occurring chiefly on alluvial grounds near streams. Its region the rainfall varies under 100b cm to 165 cm except in the Hoshangabad district, where it ranges between 175 cm to 200 cm. Teak occurs on a variety of geological formations. In some localities it attains a fair size on the Vindhyan sandstone and limestone or on alluvial ground near rivers.

The highest teak in the central India is grown is south Chanda district of Maharashtra in which the most important forests of Allapilli in the Ahivi range,are situated. Another intresting and important teak situated in a deep valley at an average elevation 435 metres the seal-level, through which runs the Bori river. Enclosed on the north by a scfarped ridge rising to 850 metres and on the south by several monor ridges rising to 570 metres. The rocks are partly traps, partly massive sandstones of the upper Gondwanas, and partly soft sandstones, clays and shells lower gondwanas.

Its occurrence in the forest areas of the study region is a common feature. Its distribution is confined more in south western and central portion of Betul plateau than that of north eastern portion. Traps have pro dominant species to almost pure form. Trap plateau and hills has the maximum concentration of the tree. It has four good abundant localities firstly, it covers the northern part of the region in Barbatpur forest range, with good quality teak. The Machna valley in the middle portion of the range has moist teak forests in percentage, here teak stands as 50 percent of the growing crop. Pure teak stands and superior quality with the height of the tree above 30 metres may be seen as in the small alluvial island in the Machna valley. Secondly, it extends in Harrai, Gowasen and Jugmadeo hills of Gawasen range. This species extend to the khokharkheda, Tamda, and kila vete where teak greatly pro dominator clay loamy soil. The quality of crop is good with well formed cylindrical trees stands river valleys of Bhangi, Dudhi and Lirbida. Thirdly, extensive teak tracts are found in the western Tapti valley ranges, particularly Mohda and Tawdi. Some times, it grows in gregarious form, on well-drained alluvial flats. On the fertile loam it may attain large dimensions. Such cases are examples of progressive succession from deciduous to an evergreen type on alluvial strips. Fourth, in Athnair range the teak grows on 60 per cent forest land. This area is trap covered high plateau and hills, here the teak is usually of small size but occurs in great abundance.

In the Betul range teak grows 30 per cent of growing crop. The rainfall in this area amount about 120 cm which causes occurrence of good quality teak specially in Baretha Ghat. The good quality of teak is mainly confined to the northern slope of Satpura range.

Table 4.4

Betul Plateau : Occurrence of Tectona grandis in the Teak Forests

Particulars	Gawasen	Tawdi	Betul	Athnair
Frequency	100.00	100.00	87.50	92.85
Relative Frequency	17.74	14.62	08.57	11.75
Density	05.72	05.30	03.42	03.76
Relative Density	17.73	14.60	09.77	12.61

Source: Based on field work.

It is evident from Table 4.4 that this tree is the predominant tree species of the forests of study area. Its presence is recorded 95 per cent of the sample quadrats. Frequency of occurrence is high in western plains of the region where it is 100 per cent in Gawasen and Tawdi forest range while the central part, shows the comparatively low frequency in Betul range. Similarly the relative frequency is high in Gawasen and low in Betul range.

Proportion of the plant in total plants occurring on sample sites is relatively very high. The highest figure 5.7 is for Gawasen and 5.3 is for Tawadi ranges. The population of this tree is comparatively low in Betul forest range (3.42) and Athnair range (3.76). Same pattern is noticed in respect of relative density.

Tectona grandis is also marked in the mixed forests. In mixed forest highest density is found in Barabatpur region and lowest in Sawalmendha range. Thus, it can be concluded that teak is more common in the mixed forests of study area.

Applied Values

Teak wood is moderately hard, extremely durable and does not wrap or split. The timber of teak is widely used for structural purposes. For ship-building its timber stands in a class by itself and has a world-wide reputation. It is extensively used for house-building, railway carriages and wagons, carpentry and numerous other purposes. The timber is used locally for making agricultural implements and also for the hubs and spokes of wheels. Its rays are very fine and have good finish so timber is of high value for furniture. It is also used for making plywood. Local people use leaves to make roof of their huts.

Terminalia tomentosa/Saj Tree

Apart from its economic value the tree is important silviculturally as being one of the commonest of Indian forest trees. Saja, a large sized deciduous tree with a well developed crown attains a height of more than 35 mts. The bark is dark grayish, with characteristic fissures resembling the skin of crocodiles and hard dark brown wood.

Leaves are coriaceous villous on the under surface, up to 18 by 8 cm, elliplicoblong, obtuse or even emerginate or slightly acute at the apex. The old leaves shed during January or February with the new leaves appearing in late May or early June. The flowering season is July. The development of fruits takes place in autumn and they ripen in the following period from March to May. Saj has a well developed and deep root system, due to which it is wind firm.

Natural Habitat

It is a tropical and sub tropical tree. It thrives in both moist and sub-moist climates and thus is a *Mesophytic* tree. The tract which experiences hot summers with the temperature being more than 40°C and winter are mild the growth of tree is well. Rainfall in the Saj occupied areas with the total annual precipitation is being more than 1000 mm. thus, the plant has occurrence from demarcation point of view in moist climate from existence point of view. Generally it is not observed beyond the rainfall boundary of 850mm. Thus the characteristic of its growth and development coincides with the eco-climatic condition of the humid type climate. It is a moderate light demander; its growth becomes suppressed under shade and sensitive to frost and drought.

Altitudinally the tree is found between 150 to 2500 metres from MSL in Indian sub continent. Plain habitat is the most suitable site for this plant where it has more density as well as frequency. Thus, it thrives in moist depressions areas i.e., valleys and adjoining slopes very well. It is the fact that soil moisture plays a vital role in determining plant behaviour. Saj grows best on rich alluvial soils with good drainage. Its growth favours stiff clayey soil, where it often becomes gregarious in depressions and round the edges of swamps. It grows freely on black cotton soil, though it remains stunted. On poor shallow soil, particularly on hilly ground, though often plentiful it remains stunted. Thus, by nature it exhibits obviously the mono climax tendency of plant succession.

Phytogeography Distribution

Terminalia tomentosa is one of the commonest and most widely distributed species of deciduous forests. It extends throughout the sub-Himalayan tract and throughout the greater part of the Indian Peninsula, extending into comparatively dry regions. In the Siwalik Hills outer Himalaya and Central India it is a common companion of the Sal as well as of the teak.

In the Himalaya it ascends to about 4,000ft. It extends westward to the Kangra hills and possibily to a small extent farther west. In the eastern part of the sub-Himalayan tract it reaches very large dimensions, for instance in the Tista forests, where it occurs both in sal forest and in low- level mixed forest without sal. In Jalpaiguri it is not plentiful, though there are some well- grown trees in the Upper Tondu forest. In Chota Nagpur the tree is common, particularly in valleys and moist localities. On village lands it is very common, and is extensively pollarded for the cultivation of the tasar silkworm.

Terminalia tomentosa is common in deciduous forests throughout the greater part of the Central Provinces and Berar; it is a common companion of the Sal as well as of the teak. It is plentiful on black cotton soil, which many species avoid, but here it is usually small. It reaches its best development on the moist fertile alluvium near rivers, where it sometimes tends to be gregarious.

Terminalia tomentosa is plentiful in most of the forest tracts of the Bombay Presidency. Being perhaps nowhere more abundant than in Kanara and in the Dangs forests of Surat. In Kanara it grows better on granite and schist than on laterite. In the Deccan it prefers black cotton soil to the shallow soils of the quartzite hills so frequent in that region.

The tree is fairly well distributed in Madras, and in some localities it is plentiful. It is characteristic of various types of mixed deciduous forest, and in Ganjam is a common companion of the sal. It is fairly plentiful in the Wynaad, and ascends to 4,200ft. in the Nilgiris. It is common in many parts of Hyderabad and Mysore, and is very common in Travancore up to 2,000ft.

The tree is found in both the compact and scattered form in the study area. In low-lying areas it is found in pure patches. It is plentiful in certain types of lower forest on flat alluvial land tending towards

1. Saj Tree (*Terminalia tomentosa*)
2. Haldu Tree (*Adina cordifolia*)

Lendia Tree (*Lagerstromia parviflora*)

dry rather than moist conditions. In teak and mixed forests this species is in associated form. At some places forests sub types are also found. The Terminalia tomentosa type sub forests is found on the northern and northeastern slopes as well as in shaded and cooled packets of the study area as below : -

(i) ***The Terminalia tomentosa***: *Ougeinia oojeinensis sub type* association is found in moist areas, along lower hill slopes of Machna valley where the soil is deep but drainage is poor. Association of this type occurs in compartment number 205 and 212.

(ii) ***The Terminalia tomentosa***: *Anogeissus latifolia sub type* association occurs on high altitude in Sarni range. The crop growth is luxuriant with saj predominating on clayey loam, where it is almost occurring as pure crop in places. In favourable localities it attains a height of over 30 metres. Good quality saj growing on flat alluvium of frost affected Tawa valley with density being uniformly good. It is also found in moist depressions where it attains a height of over 30 metres but when it occurs below its normal limit it is not of good quality and smaller in size.

The Satpura ranges and plateau, throughout the teak and mixed forests areas, it is found in scattered form. The Terminalia tomentosa forms the upper storey of the teak and mixed forests the variation may be seen in the height of trees from place to place according to depth of soil and slopes of the land. The good saj stand can be seen in the alluvium steps of regional rivers in Betul forests. In Chicloli plateau along both the sides of the Tapti valley, the alluvium are provided with moist type forests in which The Terminalia tomentosa is seen as co dominating associate species. Here The Terminalia tomentosa is scattered as small blocks in the ravines of Tapti, Batki, Labada, Khandu and its tributaries. It is plentiful on black cotton soil of Betul plateau to which many species avoid, but here it is usually small. It reaches its best development on trap plateau and is found along nalla and areas with greater soil depth, where it some time tends to be gregarious.

Table 4.5

Betul Plateau : Occurrence of Terminalia tomentosa in the Teak Forests

Particulars	Gawasen	Tawdi	Betul	Athnair
Frequency	45.45	30.76	50.00	21.42
Relative Frequency	01.97	01.27	01.78	00.72
Density	01.40	01.50	01.25	01.50
Relative Density	04.33	04.43	03.57	00.64

Source: Based on field work.

Table 4.5 makes it clear that Terminalia tomentosa species is comparatively more frequent in the central part than the southern part of the study area. The highest percentage of occurrence of frequency is marked in Betul range where its presence is recorded as 87 per cent of the quadrats but in Athnair range to the south it is found only in one fifth of the quadrats. Phytogeographical distributional pattern of the tree is also reflected by survey quadrats. On average 1.97 per cent of plant belong to this species in the Gawasen range but only 0.72 per cent in the Barbatpur range.

The density of this species is equal in almost all the forest ranges. The density is highest in Tawdi and Athnair range (1.5) followed by Gawasen range (1.4). The relative density of figure is comparatively sufficient in the region.

In mixed forests it is found during field survey that frequency of occurrence of the Saj tree is almost the same (18.7 to 14.2) in all ranges. The density is highest in Sarni (0.25) which amounts to 0.57 relative densities. Thus, density and relative density indicates the richness of the species and distributional pattern in teak and mixed forests of the study area.

Applied Values

It is a multipurpose economic plant. Near about one-third of the total interviewed respondents have reported the uses and its medicinal properties:

1. The wood of tree is strong and durable, due to this strong beneficial factor, its market value commercially at a large scale. It is used in building carts, railway wagons, for making agricultural implements and furniture.
2. The wood of the tree is not preferred as a fuel but in lack of all other alternatives it is some time used as fuel.
3. The leaves of the tree are good fodder for goats and cattle.
4. Its bark is used by villagers in ulcers.

Adina cordifolia/ Haldu Tree

It is a large deciduous tree with a large crown and cordate leaves. Under forest conditions the tree produces a long straight clean bole, but it is often butteressed and fluted at the base, the buttress sometimes being of irregular and fantastic shapes. In more isolated positions it produces a thick bole and massive branches with a large spreading crown. The tree attains large dimensions. Thus it belongs to the class of *Megaphanerophytes* from life form point of view.

Its wood is yellow, moderately hard and bark is grey exfoliating in patches. The large whitish stipules enclosing the leaf-buds are conspicuous. The leaves shed near about February and May-June is the months when the new foliage appears. From leaf-classes point of view it falls under the class of Mesophylls. The yellow globose flower heads appear from June to August. By October the globose fruit-heads are almost fully formed.

Natural Habitat

The plant has occurrence from demarcation point of view in tropical and subtropical part of India which may be termed under the sub moist climate from existence point of view. In its natural habitat the absolute maximum temperature varies from 37°C to 47°C and the absolute minimum temperature from 3°C to 10°C. The average annual rainfall varies between 75 to 350cm. Thus, the Characteristics eco-climatic condition shows the sub humid (mild or) type of climate.

The observation over the survey sites as scattered in different habitats shows the nature of distribution of Adina cordifolia. The soil requirements of the tree have already been alluded to well drain sandy loam which is essential for the best development of trees. Hence, the tree thrives well on deep boulder deposits overlain by a sufficient depth of fertile porous loam. It also grows well on alluvial ground provided with the good drainage, near rivers and on flat ground. Thus, by nature Adina cordifoia is found most frequently and attains good height along the lower slopes of hills. By nature it exhibits obviously the mono climax tendency of plant succession rather than poly climax.

Phylogeographical Distribution

Adina cordifolia is found scattered in deciduous forests throughout the greater part of India. In the sub-Himalayan tract large trees are found along the lower slopes of the outer hills from the Jumna the boulder terraces at the base of these hills, where it is sometimes almost gregarious. It is fairly common in mixed and sal forests in Chhota Nagpur and scattered throughout the greater part of Central India and Indian Peninsula generally.

It is a familiar tree in the teak and mixed forests of the area under study. In the northern part its growth is marked very well spreading over Mahadev hills, is stretched from northeast to Morand valley to the west. It is found very well in Sarni and Barbatpur ranges to the north and southern Bhanwargarh hills in the Betul range. The tree is noted in scattered patches in the forests, along the slopes of Rabinideo hill of Sarni range and Madamjhiri forests in Betul range. In the Satpura tract large trees are found along the lower slopes of the outer hills and on the boulder terraces at the base of these hills where it is some times almost gregarious. It is some times met within round the edges of swampy depressions where the soil is stiff and the drainage is deficient, but in such places it is always stunted and tends to become stage-headed early.

It is noticed as growing in Pakshi and Pat forests of Bhaura range and Kanjitalao forest of Barbatpur in mixed forest. It also registers its presence in Mahupani forest area of the Tapti valley in the central part and Morand valley in western part of study area. Adina cordifolia presence is scanty in the mixed forests of Dabka forests circle, in Sawalmendha forest range.

Table 4.6

Betul Plateau : Occurrence of Adina cordifolia in Teak Forests

Particulars	Gawasen	Tawdi	Betul	Athnair
Frequency	27.27	23.07	25.00	28.57
Relative Frequency	01..13	00.63	00.89	01.19
Density	01.33	01.00	01.50	01.25
Relative density	04.12	02.76	04.28	04.19

Source: Based on field work.

Table 4.6 reveals that the occurrence of Adina cordifolia is almost same in the teak forests of the study area. During field survey its presence is marked in more than 25 per cent of the surveyed sites. The result of the above table shows that the highest frequency (28.57) is found in Athnair range which is followed by Gawasen range. In Tawadi range towards the west, the lowest 23.07 frequency is noticed. The highest concentration of relative frequency is found same in both the ranges of Athnair and Gawasen (1.1)

The highest density and relative density occurs in Betul range and in Tawdi it is the lowest. By looking the frequency and density figures it can be concluded that it is sufficient.

During field survey it is marked in mixed forests also. Its presence is noticed only in 17 per cent on the selected study sites. In general it is a sparse species.

Applied Values

It is one of the useful plants and has multipurpose applications or rather to say it is used for the welfare of human beings. About 45 per cent respondents have informed the different uses and prescribing drug made out of Adina cordifolia.

1. Its timber is one of the important timbers particularly that of the central India. The wood is yellowish, moderately hard &

naturally durable. It can be easily and readily sawn and worked to a smooth finish, so used for building furniture, turnery and bobbins. Thus, it has good commercial value.

2. As uses and properties described by Kirtiker and Basu, parts of this plant have medicinal properties. The juice of the plant is used to kill worms in sores. The bark in antiseptic.

Dalbergia latifolia/Sisham Tree

A medium to large sized deciduous tree but may behave as evergreen, attaining a height of upto 10 to 25 metres or even more in some localities. Branches are strong but pendulous, canopy is oval or of umbrella shape. Bark is light brown to brownish, with longitudinal fissure exfoliating in strips and scales. The leaflets being orbicular and obtus or emarginated instead of rhomboidal or broad ovate and 4-7cm. long. The old leaves shed in November and December and new leaves appear in January or February. From leaf classes point of view the tree falls under the class of *Nanophylls* but from life form points of view it falls under the group of *Mesophenerophytes*. The wood is very hard, close grained, strong, durable and ornamental. It bears yellowish white flowers. Flowering season is March to April.

Natural Habitat

The tree belongs to tropical and sub-tropical regions but can not tolerate cool climate. In its natural habitat the absolute maximmum shade temperature vaies from 37°C to 47°C and the minimum from 3°C to 15°C. It prefers a part of the moist locality. The tract in which it occurs receives heavy rainfall with the total annual precipitation being more than 80cm.

The Blackwood grows on a variety of geological formations, including gneiss, trap, lateriate, boulder deposits and alluvial formations including good drainage and reaches its best development where the soil is deep and moist particularly in the neighbourhood of perennial streams. It will grow fairly well on black cotton soil.

Altitude is another essential factor in the phytogeographical distribution of this plant, so generally its occurrence is upto 1500 metres MSL. It reaches its best development on alluvial and riverine tracts. Its nature of distribution declined more towards the undulating habitat rather than that of valley or flat ground habitat respectively. Thus, by nature of plant growth as well as for survival this tree is characterized by monoclimax tendency of succession.

Phytogeographical Distribution

Although grow more or less as scattered tree in deciduous forests of sub Himalyan tract, Chhota Nagpur, Central India, Western India and Southern India. Although often found in considerable abundance the tree is not typically gregarious but grows more or less scattered in mixed deciduous forests. Exceptionally it is found gregarious in patches on alluvial flats, as in Dangs forests of Surat, Bombay. In the Western Ghats and in the fairly moist deciduous forests of southern India it attains large dimensions. In Bombay it occurs at various elevations up to 3,000ft., associated with teak, Terminalia tomentosa, and other trees and with bamboos, in mixed deciduous forests; as a rule it forms only a small percentage of the crop. It reaches its best development on deep moist soil at moderate elevations.

In Madras it occurs in many districts, but perhaps reaches its best development on the western slopes of Nilgiris up to 4,000ft., and on the Anamalais up to 4,500ft.,in mixed deciduous forests associated with teak and other trees .

In the Central Provinces it is fairly widely distributed in mixed deciduous forests, though nowhere very abundant. It attains its largest dimensions (5 to 6 ft. in girth) in Chanda. In Bihar and Orrisa the tree is somewhat scattered and does not reach large dimensions; in Chhota Nagpur it is commonest on cool aspects and along streams of Bundelkhand and adjacent tracts, and along the sub-Himalyan belt in the Ghonda and Bahraich districts, chiefly in mixed deciduous forest.

In the study area it registers its occurrence in moist depressions of the most of the forest ranges. In the northern part, the tree attains large dimensions in moist deciduous forests. It is found gregarious in patches on alluvial flats. In Amla and Betul range the tree is marked scattered on Mowar and Landi Plateau of deep soil areas. It is indigenous in the dry deciduous forest of Betul Plateau on trap, here it is usually a small tree owing to the unfavourable conditions of soil, but perhaps reaches its best development on the slopes of Tapti valley in Mohda, Tawdi and Tapti forest ranges. On Chicholi plateau, it is also found in comparatively moist situations near streams. In the southern part it is sparsely scattered in the dry deciduous forests of plateau and adjacent tracts, also along the Gawaligrah balt. On the basis of overall observations in the study area over different vegetation zones, it is concluded that the plant has no occurrence over stony and upper slope of hills but hilly ravines and alluvial patches are most suitable site for the existence of this tree where it shows more frequency as well as density.

It is planted along the roadside or in gardens by the forest department. It also shows its occurrence in agricultural waste land and near settlements.

Table 4.7

Betul Plateau : Occurrence of Dalbergia latifolia in the Teak Forests

Particulars	Gawasen	Tawedi	Betul	Athnair
Frequency	18.18	23.07	31.25	21.42
Relative Frequency	00.56	00.84	01.07	00.95
Density	01.00	01.30	01.20	01.30
Relative Density	03.09	03.58	03.43	04.36

Source: Based on field work.

Table 4.7 shows that the occurrence of the tree is registered in all selected forest ranges. In these ranges, its presence is prominent only in Betul range where the frequency is 31.25 per cent of the quadrats. The lowest frequency and relative frequency is found in Gawasen range and north western part of the study area.

In a hectare 13.00 plants of this tree were counted in Tawdi and Athnair ranges. This amounts to 3.58 and 4.36 per cent of the relative density respectively. The lowest density and relative density is marked in Gawasen range while medium density is found in Betul range.

It may be pointed out that in mixed forests areas this tree has been noticed. The highest frequency 21.42 is registered in Barbatpur range while the highest density of occurrence 1.5 in Sarni range. Hence, it can be stated that the tree is quite widely distributed in the moist deciduous forest belt but percentage of occurrence is poor on higher alleviation and dry deciduous forest in the study area.

Applied Values

It is an important multipurpose economic plant. Nearly 50 per cent respondents have reported its different uses. Its wood is strong and of excellent quality. It is a high class furniture and cabinet wood widely used. It is esteemed also for musical instruments, electric casing, hammer handles, tobacco pipes etc. It is also used for making agricultural implements. The wood of the tree is not preferred as a fuel though some times used as fuel. Its leaves are good fodder. It also has medicinal properties. A decoction of leaves is useful in gonorrhoea *(Mathur 1970, 112)*. Sisham has a dense evergreen foliage. This makes it an excellent ornamental tree for planting in gardens, parks, roadside and avenues.

Pterocarpus Marsupium/Bijasal Tree

Pterocarpus Marsupium is one of the most important timber trees of the deciduous forests of the Indian Peninsula. It is a large deciduous tree with spreading branches and a straight clean bole. Under favourable conditions it attains a height of 35 metres. Its wood is very hard, close grained & hard wood is yellowish brown. Bark is thick, grey, exfoliating in small irregular scales. Leaves are 7 to 9 inch long ovate-oblong, mucronate silky but hairy beneath. The panicles of fragrant yellow flowers appear from June to September. According to its life form it falls under the group *Mesophanerophytes*.

Natural Habitat

The climatic conditions prevailing in the tract, in which this tree occurs are subtropical In its natural habitat the absolute maximum shade temperature varies from 35°C to 48°C and the absolute minimum from 1°C to 16°C. and the normal rainfall from 75 cm to 200 cm.

The tree prefers a soil with a fair proportion of sand and is often found on red loam with a cartain percentage of clay. It appears to grow on a variety of formations provided the good drainage. It is common on gneiss, quartizite, shalles and conglomerate. It grows on rocky laterite, growth is well suited to gneissic soils but will not grow on the sandy soils.

It is found on either hilly or undulating country and on more or less flat ground. It occurs mainly on hilly ground, at elevations up to 1100 metres or sometimes more, but most commonly found between 150 to 600 metres. It reaches larger dimensions than most other trees on exposed hill sides on rocky ground, it prefers northerly aspects. Thus, by nature of plant growth as well as survival this tree species is characterized by polyclimax tendency of succession.

Phytogeographical Distribution

It is spread throughout the greater part of the Inidan Peninsula, extending northward to Mount Abu in the west, in the central India it is fairly common. The tree is found scattered in deciduous forests, mainly on hilly ground, at elevations up to 3,500 ft. or sometimes more, but most commonly between 500 and 1,500 ft. In the Central Provinces it is fairly common in Bhandara, Balaghat, Saugour, Chanda, and Damoh, and is more or less scattered or locally common in other districts, but is rare in Berar; it occurs mainly on hilly ground, preferring northerly aspects, its chief companions being teak , Terminalia tomentosa and other trees characteristic of the mixed deciduous forests. In Bombay it occurs most plentitfully in the southern portion of the Presidency, being particularly common in North Kanara and fairly plentiful in Surat. In the moist climate of North Kanara it affects chiefly the upper parts of ridges where it obtains light and warmth; its chief associates here are teak, Dalbergia Latifolia, and Adina cordifolia. In the Deccan districts it is more scattered; in the Khandesh Akrani it ascends to 3,700 ft.(Talbot). In southern India it occurs in most districts, scattered in deciduous forests associated with teak, and other trees with or without bamboos. In the Nilgiri Wynaad it may in favourable localities form as much as 10 per cent of the growing stock. It attains its largest dimensions in the south, particularly in Coimbatore, Madura, and Tinnevelly, where trees up to 16 ft in girth are occasionally met with. In the region of Pterocarpus santalinus,that is, chiefly in Cuddapah and North Arcot, the two species occur together, but whereas P. santalinus ascends to 3,000 ft. P. Marsupium ascends only to 2,500 ft. In Mysore it is common in deciduous forests. In Travancore it is common in deciduous forests up to 3,500 ft.(Bourdillon). It is one of the most important timber trees of Hyderabad. In Chota Nagpur and Orissa it is locally common, chiefly on hilly ground in valleys or on northerly slopes, in mixed deciduous forest or in Sal forest. Except in parts of Orissa, Sambalpur, and Singhbhum it is usually of small size in parts of Bundelkhand and in the sub Himalayan tract from Gorakhpur westward to the eastern corner of the Kumaun bhabar tract, in dry deciduous forest associated with Lagerstroemia parviflora trees. It is found scattered in limited quantity and of small size in parts of the Marwar, Rewah, Indore, and Gwalior forest. The tree is characteristic of deciduous types of forest, and appears to grow on a variety of formations provided the drainage is good. It grows both on hilly or undulating country, and on more or less flat ground, preferring a soil with a fair proportion of sand, and is often found on red loam with a certain percentage of clay. It is not exacting, however, since it reaches larger dimensions than most other trees on exposed hill-sides on rocky ground and shallow soil where the forest growth is poor: in such places it is at times almost gregarious. In the Peninusula it is common on gneiss, quartzite, shale, conglomerate, sandstone and laterite less common on trap (Haines). In the sub-Himalayan bhabar tract it occurs on a deep dry well-drained boulder formation where the subsoil water-level is at a great depth.

In the study area it occurs in most forests ranges, more or less scattered or locally common in teak and mixed forests associated with teak and other trees. In the region, Bijasal does not extend over as vast but still occupies an important place. The degree of variability in the spatial distribution of its species varies from place to place due to ecological adaptation.

In the Amla, Sarni and Betul forests ranges it is frequently found. On this belt soil derived from metamorphic rocks occur. These soils with greater moisture retentivity are most suitable for the growth of moist type of vegetation. Towards the slopes of Tawa valley it may be seen in favourable localities from as much as high percentage of the growing stock. It attains its largest dimensions up to 30 metres. The forests of Machna valley come next in richness of this tree. In localities where Pterocarpus Marsupium is common are northern slopes. But the southern slopes are usually poor in this species. In sandy soil areas i.e. north eastern part of the study area it does not grow. In the north western part of the study area it occurs on well drained boulder formation where the sub soil water level is at a great depth. On troppean tract it occurs less common. It is found scattered in limited quality and of small size on trap plateau and hills. It is mixed with its principal associate in low lying areas of Nanda and Gokhar forests of Chicholi range. The regional pattern of Bijasal population also shows an uneven distribution which is directly correlated to the areas they occupy.

Table 4.8

Betul Plateau : Occurrence of Pterocarpus marsupium in the Teak Forests

Particulars	Gawasen	Tawdi	Betul	Athnair
Frequency	00.00	15.38	18.75	28.57
Relative Frequency	00.00	00.63	00.53	00.96
Density	00.00	01.50	01.00	01.00
Relative Density	00.00	04.13	02.85	02.98

Source: Based on field work.

Table 4.8 indicats that Pterocarpus marsupium is not a common plant. It is not found in Gawasen range during survey. However, it is recorded in 28 per cent quadrat in Athnair range and 15 percent quadrat in Tawdi range. The highest relative frequency is also registered in Athnair (0.96) and the lowest in Betul range. The density is comparatively high in Tawdi range (1.50) and low in Betul and Athnair ranges. Same pattern is noticed in respect of relative density.

In the mixed forest it is also noticed in Tapti and Sawalmendha range. The density is marked 0.27 in Tapti and 0.37 in Sawalmendha. Hence, it can be observed that the distribution of Bijasal tree is uneven and is in scattered form in the study area.

Applied Values

Nearly 37 per cent of the total respondents have informed about its different uses. Troup (1921, 267) and Sharma and Sharma (1999, 263) have described the uses of the plant in their respective works at length.

1. The wood of tree is yellowish brown, durable & is economically valuable. It is used in agricultural implements, carts, wheel work and boats.
2. Its red gum-resion which exudes from wounds in the bark furnishes the 'kino' of commerce.
3. Its wood and gum are used in some indigenous medicines. Gum is valuable astringent used in medicine and wood is also a medicine for diabetes.

Emblica officinalis/Aonla Tree

It is a small or middle sized tree with crooked trunk and spreading branch. Branchlets are glabrous or finely pubescent on which leaves are closely set, like a fan, olive shaped, 1-2 cm long and 0.5 cm broad. Flowers are small, light green to pale green in colour and appear in clusters in spring. Fruit is pale green, fleshy globose or round, light with 5 to 6 cm in diametre. The plant belongs to the class of *Microphenerophytes* some time *Mesophanerophytes* as a tree and as for its leaf size, it belongs to class of *Leptophylls.*

Natural Habitat

The climatic conditions prevailing in the tract in which this tree occurs in the region are warm and humid climatic zone. In summer the temperature may soar to over 40°C, while winter are mild. The total annual precipitation is more than 75 cm.

The tree is found most frequently on well-drained ground, for instance along the lower slopes of hills. It also grows well in the valleys where the drainage is good.

The tree accommodates itself to a variety of soils and geological formation, including black cotton soil and trap. It thrives best on deep porous loam.

Phytogeographical Distribution

The plant grows throughout tropical and subtropical regions of India. It was found to be abundant in deciduous forests of Central India.

Emblica officinalis is one of the most common associate species of mixed and teak forests of the study area. It is distributed throughout the area in scattered form, although in the western part it is more numberous. Here, it is found in compact patches particularly in the Tokra forest of Saoligarh range, Morand valley in Chicholi range and catchment areas of Landtha in Tawadi range. It is marked abundantly in Kukru forests of Bhainsdehi range and Pothakund circle of Sawalmendha range to the south. Chopna forests of Sarni range and Kajlitara of Barbatpur range to the north, show its occurrence. The pure crop of the tree is not marked in the area under study. During general observation, it is noticed that Emblica officinalis is frequently seen in black soil areas of the teak forests. In the area with good soil depth, natural regeneration of Emblica officinalis has been observed. It is planted on the outskirts of the agricultural field.

Table 4.9

Betul Plateau : Occurrence of Emblica officinalis in the Teak Forests

Particulars	Gawasen	Tawdi	Betul	Athnair
Frequency	45.45	30.76	37.50	14.28
Relative Frequency	01.69	01.06	01.07	00.72
Density	01.60	01.50	01.20	01.50
Relative Density	04.95	04.13	03.42	05.03

Source: Based on field work.

The Emblica officinalis tree is registered one fourth of the quadrats in all the surveyed sites of teak forest area. The results of the survey present in Table 4.9 shows that the highest occurrence is marked in Gawasen (45.45). The relative frequency is also high in this range. It was noted during observation that the tree is more numberous in deep soil areas. In the lowest frequency 14 per cent of occurrence is in Athnair range. Here, relative frequency is also the lowest.

The individual population of the Emblica officinalis is highly concentrated in Gawasen (1.60) and the lowest in Betul (1.20). The relative density is the highest in Athnair (5.03). The relative density figure is comparatively sufficient in the study area. During field survey this species is also noticed in mixed forest.

Table 4.10

Betul Plateau : Occurrence of Emblica officinalis in Mixed Forests

Particulars	Sarni	Barbatpur	Tapti	Sawalmendha
Frequency	25.00	28.57	16.66	31.25
Relative Frequency	00.72	01.29	00.64	01.10
Density	00.39	00.43	00.27	00.50
Relative Density	00.64	00.83	00.71	01.14

Source: Based on field work.

Table 4.10 elucidates that the best localities of the Emblica officinalis are found in the southern part of the region. The increasing frequency from central to southern and northern part supports the above statement. The highest percentage (31.25) of occurrence is found in Sawalmendha forest range while the lowest frequency of occurrence (16.66) is in Tapti range. But the highest relative frequency (1.29) is registered in Barbatpur range whereas the lowest figure (0.64) is for Tapti range.

The highest concentration of density is found in Sawalmendha (0.50) which amounts to the highest relative density (1.14) but the lowest relative density (0.64) is marked in Sarni range. Therefore it is found in good number in the mixed forests of Sawalmendha range.

During Aonla collection season (January-February), it is very common site to see trees are badly hacked and lopped in the natural

Siris Tree (*Albizzia procera*)

Aonla Tree (*Emblica officinalis*)

forests. Measures to conserve and propagate Aonla tree is therefore urgently needed.

Applied Values

The plant ranks first in order of its uses. About 82 per cent of the total respondents (Table 4.1) have reported making its use for medicinal purpose. The fruit is one of the components of the well known Ayurvedic preparation 'Triphala'. To juxtapose with the field survey the uses and properties are reported by Kirtikar and Basu (1933, 2220-2222) Chopra and others (1956, 106) and Sinha (1996, 120-124) are as follows.

The fruits are a good liver tonic, cooling and laxative useful in indigestion, vomiting, urinary discharges, anaemia, jaundice and certain heart complaints. It is a very rich source of vitamin 'C'. The leaf is used in ophthalmia and incipient blindness and the seed is useful in leucorrhoea, asthma and biliousness. Dried fruits are useful in diarrhoea. The exudation from the incisions on the fruits is used as an external application in inflammation of the eye.

This tree is extremely cultivated in wastelands by farmers. It is planted as an ornamental tree in gardens, parks and in temples as religious tree. Leaves are used as manure, which help in improvement of alkali soils. Its fruits are used to make herbal shampoos.

Cassia fistula/Amaltas Tree

It is a moderate sized deciduous tree that attains a height of about 10 mts. with clear bole of nearly half its length. Branches are strong and slightly bent upward. Bark is light to dark greenish brown in colour and 5-6cm. long, acute or obtuse. The tree sheds its leaves during early summer (March-May). The new leaves appearing in April-May, these are of beautiful rich copper colour. Bright yellow flowers in very large hanging bunches, appear from April to June. Fruit is a pad, long, cylindrical to round and dark brown.

Natural Habitat

Cassia fistula thrives in tropical and sub tropical climate zones. It shows a wide range of climatic requirements. In its natural habitat the maximum temperature varies from 37°C to 49°C the minimum from 15°C to 18°C and the normal rainfall from 48 to 275 cm. It survies well on both higher elevation and low lying areas. It is found on a variety of geological formations. Its best growth is attained on sandstone, quartzite, granite and schist with an overlying soil of sandy loam. On such formation the tree frequently seen with large size. On trap formation characterized by shallow soil, it does not reach a height and is found scattered.

Phytogeographical Distribution

It is common in deciduous forests throughout the greater part of India ascending to 4000 ft. in the Himalaya.

It is found throughout the forest area of the region. It is one of the common associate species of the under storey of both the mixed and teak forests. In the study area, its occurrence is noted up to an elevation of 780 metres on Satpura range. It is also marked growing in the plant ecology of hilly areas in stunted growth of tree and poor density. Though the pure crop of this tree is not marked but on lower slopes and low lying areas its growth is good in number. It is more frequently found in the belts of mixed forests along the Tapti River. It is also noticed in the roadside forests especially moist type teak forests on Sarni-Damua (Chhindwara) road in Sarni range. Munda and Mendakhera forest areas in Barbatpur range also show its occurrence. Other localities where the growth is noticed are mixed forests of Amla and Sawalmendha range. It is found scanty in dry type teak forests of Gawasen and Mohda range. It is planted too along the roadside or in gardens by the forest department.

Table 4.11

Betul Plateau : Occurrence of Cassia fistula in the Teak Forests

Particulars	Gawasen	Tawdi	Betul	Athnair
Frequency	18.18	15.38	31.25	21.42
Relative Frequency	00.84	00.84	01.25	01.19
Density	01.50	02.00	01.40	01.66
Relative Density	04.63	05.50	04.00	05.59

Source: Based on field work.

Table 4.11 makes it clear that the presence of Cassia fistula is noticed in nearly one third of quadrates in Betul range. This range located in the central part of the study area more than one fifth quadrats are marked in the southern part i.e. Athnair range. The lowest percentage is seen in Gawasen in the more surveyed quadrats. The density per quadrat is highly registered in Tawdi (2.00) and lowest in Betul range (1.40). It can be observed from these accounts that the distribution of Cassia fistula is of the scattered type in the teak forests of the region. The only exception is the Tawdi area where fairly density is relatively high.

Table 4.12

Betul Plateau : Occurrence of Cassia fistula in the Mixed Forests.

Particulars	Sarni	Barbatpur	Tapti	Sawalmendha
Frequency	27.77	28.57	38.88	25.00
Relative Frequency	00.79	01.29	01.49	00.88
Density	00.44	00.50	00.55	00.25
Relative Density	00.72	00.96	01.45	00.57

Source: Based on field work.

Table 4.12 narrates the highest percentage (38.88) of occurrence of the tree in the central part of the region i.e. Tapti range. Barbatpur range stands second with 28.57 per cent. The lowest frequency is noticed in Tapti range (1.49). Thus its presence is noted in about two fifth to one fourth of the surveyed quadrats.

The density is comparatively high in Tapti forest range (0.55) and low in Sawalmendha range (0.25), same pattern is noticed with respect to relative density. Hence it can be concluded that the occurrence as well as density of Amaltas tree is frequent and high in the central section.

Applied Values

The tree stands eighth position in order of its uses. As nearly 54.7 per cent of respondents reported its use or its parts as medicine. Babbar and other (1979, 452), D'Souza (1994, 336-339), Laxmi and Kolamal (1974, 68-61) have also described the uses and properties in their respective work. Root of the plant is useful in skin disease, leprosy and tuberculous glands. Leaves heal ulcers and used in rheumatism. Pods are crushed to make a paste which is used to cure intestinal worms in infants. Herbal vendors use Amaltas for the treatment of constipation and as an anthelmintic. The pulp from the fruits called cassia pulp is a well known laxative. In large quantity it causes purging and griping. Modern researches have showed it to be a weapon of great anti-viral activity, inhibiting several kinds of viruses. It has interferon like activity in destroying the viruses.

The bark of this tree is known as 'Sumari' and is rich in tannins. The timber of the tree is strong, used for making domestic articles. Since this plant species is beautiful it is preferred for planting along roads, avenue, and in parks and gardens.

Bamboos

Bamboos are not trees in the strictest sense, a group of tall arborescent grasses. They are medium sized perennial species attaining a height of up to 15mts. or even more in some case. Skin grayish green, of ten blotched, close packed in dense clums. Bamboos grow in clums which comprise of numberious clums. A series of internodes occurring at intervals, with solid, transvers walls separating them. The clumps of most species are erect. The rhizomes give off now culms mainly in the rainy season. Its structure resembles that of the culms lying above ground. The true roots of bamboos develop from the nodes of the stem. Leaves densely clothed on both sides with deciduous hairs, linear 10-20 cm. long and narrow. Flowering is irregular.

Natural Habitat

In the study area, it can be observed by the survey of forests that bamboos are result of the interactions between climatic, edaphic and physiographic factors. The tract in which this species occurs receives total annual rainfall of more than 750 mm. and the temperature may soar to over 40°C. in summer. Thus the characteristics of its growth and development coincide with the eco climatic condition of the semi dry to moist type climate.

Bamboos grow on different types of soils, varying from clayey to coarse soils. It requires a very well drained soil which should be deep and fairly porous. The dense bamboos are found in the alluvium of river valleys and on gentle slopes with deep fertile loamy soils.

The average range of its altitudinal distribution lies in between 300 to 1500 metres MSL Again this altitudinal limit varies from locality to locality, being higher in southern than in the northern slopes. The pure bamboos occur naturally on lower slopes of hills, slopes along nalas, ravines and in the valleys at suitable elevation. The plant is generally absent or shows no occurrence on ridges, flat top of hills and dry upper slopes to the southern aspect. Thus, by nature, it shows some or more mono climax and lesser polyclimax tendency of plant succession.

Phytogeographical Distribution

In the study area bamboo is found scattered among the species of teak and mixed forests as associate. Pure stands of bamboo are found only in small patches. Bamboo forests are good and at places form pure bamboo forests, referred to as 'Bamboo breaks' Extensive bamboo areas lie in Gawasen, Saoligarh, Barbatpur, Betul and Amla forests ranges. In these areas bamboo thrives very well because of good soil derived from crystclline rocks. The proportion of bamboo is high in the north division. In this division, they extend over 44111 hectares land. Bamboo growth is very dense in a few compartment of Gawasen, Barbatpur and Betul ranges. In Asir range it is found scattered in small units only. In parts of Asir range and adjoining Bori reserve solid bamboo culm can be obtained. The dense bamboo forests are found in the alluvium of river valleys while on humid slopes they occur along with the species of teak and mixed forests. Its species grow best and reaches its largest dimensions on northern slopes of Satpura ranges. It also occurs on southern slopes but here the growth is usually poor and young plant can as a rule be started only with the aid shelters.

The south division is characterized by very low proportion of bamboo on undulating and eroded land part. In certain localities as in Bakur and Lonia forests of Amla range bamboo grows so densely that the area can be classed as bamboo forests with scattered over wood of trees. Bamboo very often forms impenetrable masses. It is found only along the sheltered valleys. Generally speaking, bamboo grows best on slopes of hills and along ravines and nala valleys. On ridges and flat top of hills it disappears. It also avoids dry upper slopes to the southern aspect. The dense bamboo forests are found in the alluvium of river valleys while on humid slopes they occur along with the species of teak and mixed forests. The bamboos are also found scattered in alluvium and ravines of the Tapti River of the Tapti range in the south division. In the west division bamboo occurs in the river valleys and lower slopes of Saoligarh hills.

Table 4.13

Betul Plateau : Occurrence of Bamboos in the Teak and Mixed Forests

Teak forest Range	Frequency	Mixed forest range	Frequency
Gawasen	26.08	Sarni	33.33
Tawdi	53.84	Barbatpur	28.50
Betul	50.00	Tapti	22.22
Athnair	35.71	Sawalmendha	18.75

Source: Based on field survey.

The Bamboo is frequently noticed in all the ranges surveyed. Table 4.13 shows that in Tawdi range it occurs in more than one half quadrats. It has registered a frequency of 53.8 per cent. In respect of frequency, Tawdi is at the top followed by Betul range. In case of Gawasen range it appears in slightly more than one fourth of quadrats but show low frequency (26.0) in the range.

The number of this species is uncountable so their density has not been determined.

In the study area its occurrence is also noticed in the mixed forests. It is clear from table 4.13 given above that this species is occurring in one third quadrats of the mixed forests under survey and frequency is the highest in Sarni and the lowest in Sawalmendha range.

In the area under study no part of the ranges is managed exclusively as bamboo forests. Bamboo continues to be treated as an under growth. They play a very significant ecological role in the regeneration of teak. But rapid rate of cutting of Bamboo forests has led to the disappearance of the species. Bamboo is being supplied to the Nepanagar Paper Mill since a long back by which the deforestation takes place. Every year fires slowing down increment of the bamboo rhizomes. The tribals of the area totally depend upon forests. They cut bamboo to make articles like baskets, toys and many other objects for their livelihood. Over exploitation of bamboo in accessible areas near villages are largely responsible for the receding of bamboo areas surrounding inhabitations, especially in smaller isolated protected forests. Thus it can be concluded that this species has reached to the sub-climax vegetation type.

Applied Values

Bamboos are perhaps the most widely used plant species and there are many uses of bamboos as below:

1. The people living in the rural areas have been using bamboos for constructing hutments and houses. The Baskets, Mats, Ropes made from bamboo are widely used. Furniture made from bamboos is very popular in urban centers. It is used for agriculture and many commercial purposes.
2. Bamboo's green leaves and young culms are preferred as fodder by cattle.
3. It is the best quality raw material for manufacturing of pulp and paper.
4. They may be planted for ornamental purposes, soil moisture conservation and ecological stability.

ENVIRONMENTAL IMPACT ASSESSMENT

Teak grows by dint of nature and the study area consists of most valuable teak forests. Betul teak is famous all over the country and compared well with the Burma teak. The crop is a resultant of climatic, biotic and edaphic factors. The climatic conditions of the plateau are warm and moist which is suitable for the growth of teak.

The moist teak forests occur on metamorphic and crystalline rocks with a lower percentage of crops than in dry types occurring on trap with cylindrical and well formed boles. Dry type teak forests occurs in major parts of the study area; the forests are being mainly associated with trap where teak existing as almost pure crop, almost to the exclusion of other species but teak is poorer in form and girth to those growing in moist zone.

In the study area an average individual population of teak in a hectare accounts 192 trees. The western part of the region is highly concentrated in teak population where 53.5 per cent is teak of growing crop. On account of distribution of both the types, the density is not uniform. The density of moist type is usually good and may vary from 0.7 to 0.9 or occasionally a full canopy may be seen. It is highest on flat alluvial patches or on lower slopes of sheltered valleys and on humid slopes. The dry types have an average density of 0.5 to 0.7 with large blanks usually being found in the midst of well stocked forests which are the results of the shifting cultivation in the past. This type of forests occurs on higher elevation, exposed, slopes, steeply racing residual hills, and eroded areas and on all aspects, in short under greatly varying conditions.

There is a preponderance of middle aged trees. Mature trees are lacking on account of their successive removal. Regeneration of teak is patchy and inadequate. Natural regeneration of teak occurs in scattered patches especially in well drained areas.

Soils produced by the Deccan trap favour teak often to the almost exclusion of all other species though not growing to a very big size on account of shallow soil depth. However, formations of lateritic soil are high on flat grounds which are not adequately covered by vegetation. The quality of crop is good in the sheltered valleys and area with deep soil. Small patches of good quality occur in the Machna valley. The area with trap is being converted into bad land due to erosion, therefore the teak area is also being reduced.

The common injuries to which the crop is subjected are fire, grazing, illicit felling etc. By these activities teak is eliminated and the proportion of Xerophyte species is increasing. The grazing pressure is unevenly distributed. Heavy grazing is restricted to the peripheral forest areas. The area under teak is being resulted in the river valleys and alluvium plains by the tribals for agriculture. Teak is thus in a seral stage and unless these adverse fellings are controlled, the growth of miscellaneous species may appear to be the climax for the region.

CONCLUSION

In the study area, it can be concluded that teak forests are the result of the interaction between edaphic, climatic and physiographic factors. Ecologically teak forests are grouped into moist and dry type of deciduous nature. The distribution of teak forests is determined by the parent rocks from which the soil in derived. The moist teak forests are found on the alluvial strips of moist valleys and crystalline rocks with gentle slopes and sheltered aspects. Soil derived from Deccan trap favours teak, thus it is intimately associated and thrives exclusively on the trap plateau and hills. The proportion of teak in the over wood varies from 30 to 100 per cent. The most familier features of teak forests are the occurrence of Terminalia tomentosa (Saj),Dalbergia latifolia(Sisham),Pterocarpus marsuprum (Bijasal),Adina cordifolia (Haldu),Ougeinia oojeinensis(Tinsa) and bamboos. The density of moist type is usually good and may vary from 0.7 to 0.9 while dry type has an average density of 0.5 to 0.7.

Most of the associate species of teak forests are used for building, furniture, commercial and industrial purposes.As Tectona grandis(teak) is the most precious wood, extremely durable with good finish so its timber is the first choice of the people for use, therefore, it is of high economic value. Dalbergia latifolia is the next precious wood due to workably hard resistant qualities. Pterocarpus marsuprum produces a good timber. Emblica officinalis fruits are one of the well known ayurvadic medicine and is a very rich source of vitamin 'c'. Cassia fistula tree is also indigenous medicinal and ornamental plant species.

Teak is spread over about 49 per cent forest area and cover 19 per cent area of the total geographical area. The teak forests occur in all the forest ranges of the region. The maximum proportion of about 79 per cent teak is found in the north division, whereas the lowest percentage is in the south division with nearly 53 per cent. According to the ecological point of view these forests along with various plant species, are in seral stage and their ecological status is termed as Climax-type. Since then, the landscape has changed enormously because of erosion, excessive cutting, overgrazing, repeated fires and human interferences. On the basis of above observation it is concluded that teak forests are of biotic-climax type.

REFERENCES

Chopra, R.N. Nayar, S.L. and Chopra, I.C., 1956, *Glossary of Indian Medicinal Plants,* New Delhi, Council of scientific and Industrial Research.

Kirtikar, K.R. and Basu, B.D., 1933, *Indian Medicinal Plants,* Second Edition, Dehradun, International Book Distributors.

Mathur, R.C., 1970, *Systematic Botany*, Agra, Agra Book store.

Sinha, Rajiv, K., 1996, *Ethnobotany,* The Renaissance of Traditional Herbal Medicine, Jaipur, Ina-Shree Publishers.

Troup, R.S., 1921, *The Silviculture of Indian Trees,* Oxford, Clarendon Press.

Sharma, V.K., and Sharma, V.D., 1999, *Prosperity Through Trees,* New Delhi, A.P.H. Pub. Corporation.

Champion, H.G., 1936, A preliminary study of forests types of India and Burma, *Indian Forest Research Journal,* Vol. I(i), 1-286.

Babber, O.P., et al., 1979, Occurrence of Interferon Like anti-viral and anti-tumour factors in extracts of same indigenous plants, *Indian Journal of Expt.,* Bot. Vol. 17, 451-454.

D'Souza, Marie 1994, *Tribal Medicine,* New Delhi, Society for promotion of waste land Development.

Lakshmi, N. and Kolammal, M., 1974, Pharmacognostic studies on Cassia fistula Linn., *The Journal of Research in Indian Medicine,* Vol. 9(3), 59-81.

5

Ecological Status of Mixed Forests

Satpura on account of the diverse topography and climatic extremes, obviously supports a much varied and complex physical characteristics and has provided the store of immense forest wealth. A major part of the Satpura plateau is covered with forests which constitute an enormous wealth of the region. Most physical habitat are suitable for occupation by many different type of plant but a particular type of plant species may exist in a given area on account of suitable environment *(Sharma 1988, 95)*. The mixed forests are composed of trees of different species in same canopy. The species composing the mixture is usually distinguished on the basis of economic importance with scattered patches of densely growing grasses intermixed with shrubs.

The mixed forest resources in the Betul plateau play a central economic role in soil and water conservation and act as a nutrient bank of basic economic activities of agriculture and animal husbandry. They form basis for energy and nutrient flow to the live stock and into agriculture. Further, the demand of domestic energy and output environment are met from the forest, therefore, forest has the key role in the energetic of village ecosystem. Wood cut directly from the forests for fuel is usually collected as needed.

The mixed forests of the study area are tremendously rich in diverse herbal medicinal resources where these plants grow as an associate species. The area under study is properly developed with 80 per cent of population which still depends on indigenous herbal medicines.

Some of the benefits from the forests thus be evaluated in Rupees, others cannot but are none the less important. On the one hand they are the source of every possible kind of material values, industrial raw materials and many bye-products. Thus, the forests of the plateau are not only important for yielding large volume of commercial timber but are also known for number of commercially useful plant. These forest products, other than timber, are referred to as Minor Forest Products. These M.F.P. apart from being valuable, give employment to large number of people particularly to the large section of population (Korku and Gond) living in and around forest areas who get part time jobs in the collection of minor forests products.

According to the Champion's classification there are two types of mixed forests occured in the study area i.e. moist and dry type mixed forests of deciduous nature. These types of forests mainly classified on the basis of Eco-climate and habitat characteristics. So these two factors are described as follows:

Eco-climate

The climatic condition prevailing in the tract, in which this forest occurs, ranges from tropical to sub tropical. Thus experiences very hot summers and the temperature may soar to over 40°C with mean annual temperature ranges from 24°C to 29°C. The heavy rains are received in the monsoon season, with a part of the total annual precipitation also being received in the winter months. The average annual rainfall varies between 75cm. to 190cm. The number of rainy days is from 40 to 85. In the study area the mean annual temperature is 24.5°C. The highest maximum temperature ranges from 28.6°C. to 42.7°C. The normal ranges of annual rainfall are 76.8cm. to 158cm. with mean annual rainfall of 117cm. The average number of rainy days in a year is 59 days. Determinations of the reliability of rainfall occurrence, for individual months are shown below in the table.

Table 5.1

Betul Plateau : Reliability of Rainfall

Rainy season	All months reliably rainy
Post-rains season	Often rainy
Early cool season	Generally rainy
Middle part of cool season	Usually dry
Later part of cool season	Generally rainy
Summer season	Usually dry

In a year the most effective period of plant growth is 6 months. Such conditions are ordinarily regarded as suitable for the occurrence of deciduous vegetation of good growth. Accordingly the natural vegetation is mostly of dry deciduous type and occasionally of moist deciduous type in area with rainfall above 125cm.

Habitat Characteristics

In Central India or peninsular region the mixed forests occur on a variety of geological formation. These forests are practically speaking, confined to the Gondwana formation, consisting of earthy sandstones. Soil derived from sandstone of infertile sandy soils, has been occupied mainly by miscellaneous species. Soils produced from metamorphic rocks are heavier and compact in structure as compared to trap soils and so less prone to erosion. These heavy soils with greater moisture retentivity are more suitable for miscellaneous species. Soils with impeded drainage on account of presence of phyllites or schists or along the nala banks and adjoining slopes are occupied by miscellaneous species.

Dry mixed deciduous forests usually occupy better aspects but with poor and shallow soil. While moist mixed forests occur on gravelly sandy loam to clay with depth from moderate to very deep. These forests are common in hilly, undulating country, denuded or hard ground. Upper mixed forests which are found on hilly or undulating terrain are of dry type than of the moist type. Lower mixed forests on flat alluvial land are tending towards dry rather than moist conditions. In the Indian peninsula they occur most frequently in moist valleys and lower slopes, where they contain mixed forests of decidedly good quality.

DISTRIBUTION OF MIXED FOREST

Mixed forests are found in all the forest ranges of the study area. The highest percentage of these forests is found in the north division. It consists of 50 per cent mixed forests of the total mixed forest area in the plateau and 27 per cent of the total forest area of the north division. The lowest percentage of these forests is in the west division. In the west division they spread over 4.4 per cent of the total forested area. The distribution of mixed forests is shown in the Table 5.2.

Mixed forests are extended over northern spur of the Mahadev hills, the Khamla plateau to the south and southern slope of the Gawaligarh hills. These forests are also distributed over the southern parts of the Betul and Amla range, eastern part of the Tapti range, northern part of the Bhaura range, south-west part of the Bhainsdehi range, southern part of the Athnair range and almost the whole of the Asir hills (Rampur-Bhatodi Project). They are also found in patches on all the ranges where conditions are unfavourable for the growth of pure teak.

Mixed forest of dry type is mainly influenced by underlying rocks which is Gondwana sandstone and occurs on high altitude of metamorphic rocks. The moist ravines and deep soil in valleys is mainly responsible for the growth of moist type mixed forests. The density of these forests varies between 5 to 7 according to the nature of soil. The density is low in the area with the schists rocks, lying just below the ground surface. The north-east portion of the study area consists of sub-tropical wet mixed forest which is spread over high hill tops of the Mahadev hills (Keelandeo and Paraskot, Matyardeo and Sirrikot).

The proportion of area under mixed forests is from 0.2 per cent in Mohda forest range to 49.45 per cent is Sarni range. Considering the range–wise distribution of area under mixed forests, the forests

Table 5.2

Betul Plateau : Area under Mixed Forests

Division	Total area under forests (in hect.)	Total area under mixed forests (in hect.)	Mixed forests area as per cent of total area	Percentage of regional area under mixed
North Division	1,23,215.80	33,013.00	26.79	50.12
South Division	1,57,078.00	20,434.80	13.00	31.03
West Division	93,112.00	2,870.00	03.00	04.35
Rampur-Bhatodi Project	30,729.50	9,546.50	31.06	14.49

Source: The conservator of Forests, Betul 1998.

ranges have been arranged into five categories viz; high, medium high, medium, medium low, and low. The low proportion of mixed forests is in Mohda, Multai and Tawdi range where the percentage of these forests is less than 1.0, while it is high in Sarni and Amla forest ranges (Plate 5.1). The details are given in the Table 5.3.

Mixed forests of the study area can be studied by grouping them in the following ranges;

1. The Sarni Range.
2. The Asir Range. (Rampur-Bhatodi Project)
3. The Bhaura Range.
4. Other Range.

The Sarni Range

The forests of the range are spread over the southern part of the Mahadev hills to the north-east part of the study area. They cover an area of 299.3sq kilometres. The whole range consists of moist mixed forests. They are extended over the Keelandeo hills, moist plateau of Landi and in the valley of Tawa and its tributaries. In the valleys they are dense.

The density of the crop ranges between 7 to 8. The ravines of Tawa valley are densely forested as they are very far away from the interference of human being. In hilly areas with sandstone soil, the density is found between 4 to 5.

Rampur- Bhatodi Project

Mixed forests of dry nature occur on 31 per cent land of the project area. In the region they are extended over Asir hill of the Mahadev hills, Bamandeo, Salimal Pahar, the escarpment of the other hills and along the Tawa and its

tributaries. The area consists sandstone of the Gondwana system which causes poor quality of forests. The trees are stunted and do not attend a large size. In hilly areas the density falls up to 0.3 because of erosion, mineral deficiency and the presence of schist rocks just below the surface of water logging during the rains.

PLATE 5.1

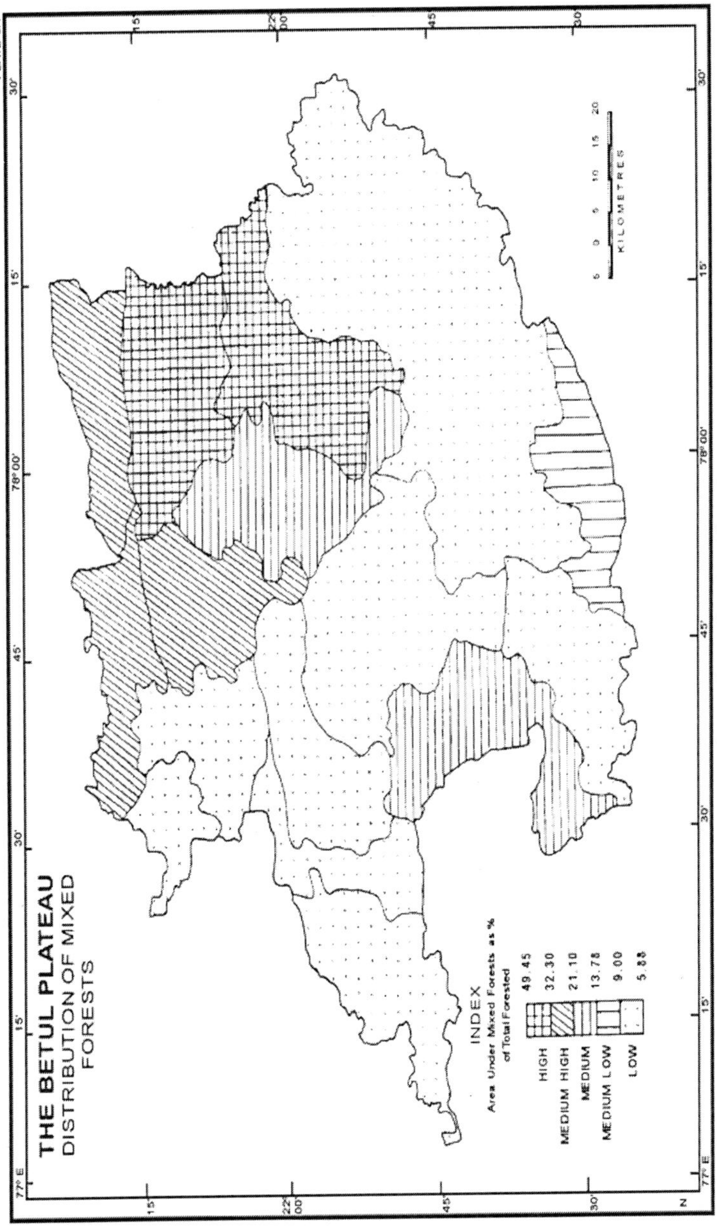

Source : Foreest Department Records

Table 5.8

Betul Plateau : Forest Range-Wise Area under Mixed forests

No.	Range	Total area of the range (in hect.)	Mixed Forests area (in hect.)	Percentage of mixed forest area
1.	Mohda	32,049.00	53.00	00.17
2.	Multai	23,765.00	85.00	00.35
3.	Tawdi	25,045.00	228.00	00.91
4.	Gawasen	16,508.90	357.10	02.16
5.	Sawalmendha	28,925.00	1,471.43	05.08
6.	Saoligarh	15,351.00	808.00	05.26
7.	Tapti	25,441.00	1,968.86	07.73
8.	Chicholi	20,667.00	1,781.00	08.61
9.	Athnair	27,394.00	3,605.00	13.16
10.	Bhainsdehi	27,752.00	4,315.50	15.55
11.	Betul	21,659.70	3,605.00	16.64
12.	Bhaura	29,416.50	7,585.70	25.79
13.	Barbatpur	25,704.10	6,666.80	25.93
14.	Amla	23,801.00	8,988.97	37.76
15.	Sarni	29,926.60	14,800.40	49.45
16.	Rampur-Bhatodi Project	30,729.50	9,546.55	31.06

Source: Compiled from records of the office of the conservator of forests, Betul 1998.

The Bhaura Range

The mixed forests of the Bhaura range are found on the northern boundary of the study area particularly in the narrow belt. They cover 7,585.7 hectares area of the range and consist 25.8 per cent forests of the total forest area of the range. These forests occur on sandstone formation around Khapa and Lonia. The poor quality mixed forests are found around Deothan forest village of the range with density varying from 0.5 to 0.7. Forests of the Bhaura range consist mainly of pollarded stems of Bhirra species.

Other Ranges

The Bhainsdehi range of the south division is covered with mixed forests. It covers 16 per cent land of the total mixed forests area and spreads over 4315.5 hectares area. In the range the mixed forests are found in the high land of the Khamla plateau to the south. The Athnair range is spread over 3605 hectares and covers 14 per cent land of the total forest area. In this range the forests are extended over southern slope of the Gawaligarh hills. The Chikhalajhiri hills, south-west and southern slope of Dhar hills also fall under the forest of the Athnair range. From the east to west of the study area, mixed forests are found as long narrow belt to the northern escarpment of the plateau.

The Barbatpur range consists of 25 per cent area under mixed forests which is extended over southern Bhanwargarh hills in patches. The exposed hill tops of the Betul range consist of mixed forests in 16.64 per cent land of the range. In Amla range, forests are found in Paraskot hills and high plateau areas while in the Tapti range they are extended over undulating valleys and in ravines.

The west division consists of mixed forests in patches along with the teak forests.

Stratification of the Mixed Forests

A very large number of species make up the growing stock. Growth of trees is in stunt form and mainly consists of non-timber miscellaneous species. Except Terminalia tomentosa (Saj), Anogeissus latifolia (Dhaura) and Chloroxylon swietenia (Bhirra). None of the species are so gregarious as to form pure crop. Mixed forests have luxuriant growth with all stories.

The top canopy is formed by the large trees of miscellaneous species along with Tectona grandis (Teak), Terminalia tomentosa,

Agrogeissus latifolia and Chloroxylon surietenia. The main species of upper storey are Terminalia tomentosa (Saj), Legerstroemia parviflora (Lendia), Diasporous melonoxylon (Tendu), Madhuca latifolia (Mahua), Terminaila bellirica (Beheda), Mitraguna parviflora (Kalamb), Adina cordifolia (Haldu), Bombox ceiba (Semul), Schrebera swietinioides (Mokha), Albizzia odoratissima (Chichwa), Soyamida febrifuga (Rohna), Schleichera oleosa (Kusum), Pterocarpus marsupium (Bija), Stereospermum suaveolens (Pader), Lennea grandis (Moin), Garuga pinnata (Keker), Dalbergia paniculata (Pangra), Grewia hirsuta (Siwan), Ailanthus excelsa (Maharukh). In moist localities Terminalia arjuna (Koha), Syzygium cumini (Jamun), Ficus glomerata (Gular), Pongamia pirnata (Karanji), are found.

The middle storey consists of Ougenia oojeinensis (Tinsa), Emblica officinalis (Aonla), Buchanania lanzan (Achar), Grewia tiliaefolia (Dhaman), Terminalia chebula (Harra), Miliusa tomentosa (Kari), Cassia fistula (Amaltas), Balanites aegyotiaca (Hingan), Semecarpus anacardium (Bhiwan), Acaeia catechu (Kher), Acacia leucophloea (Reunjha) and Aegla marmelos (Bel). A few trees of Hardwickia binata (Anjan) grow scattered in the south-western part of the Asir range. Undergrowth consists of Vifex negundo (Nirgud), Colebrookia oppositifolia (Bhandar), Holarrhena antidysenterica (Dudhi) etc.

The ground storey is formed by grasses which consist of Sehima nervosum (Sainar), Heteropogon contortus (Kusal), Cymbopogon martini (Rasha), Vetiveria zizanioides (Khus), Iscilema laxum (Muchbel), Saccherum spontaneum (Kans), and Eulaliopsia binata (Sabai).

Bauhinia vahlii (Mahul), Butea superba (Palasbal), Zizyphus rugosa (Ehurni), Milletia auriculata (Gurar) and Smilax macrophylla (Ramdatun) are the species fall under climbers category and play an important role in forming the top canopy.

VALUABLE SPECIES

Lagerstroemia parvifolia/Landia Tree

It is a medium to large sized deciduous tree with a well developed

Betul Plateau : Mixed Forests

Betul Plateau : Arjun Tree (*Terminalia arjuna*)

crown. It attains a height of more than 25 metres. Its bark is light grey to reddish, than smooth, exfoliating in narrow longitudinal flaks light brown inside. The leaves turn brown towards the end of the cold season and are shed in February-March. The new leaves start appearing in April-May. The panicles of small white fragrant flowers appear from April to June. The ripening of fruits takes place in winter, when the capsules open and the seeds are released. From life forms point of view the tree falls under the group of *Mesophanerophytes*.

Natural Habitat

Lagerstroemia parvifolia is found in different bio-climate zone in India. In its natural habitat the mean annual temperature ranges from 24°C to 30°C. The tract in which this tree occurs experience a very hot summer and mild winter and where frost is experienced in the winter season. The annual precipitation within its natural habitat should be between 750 to 2500mm. It is a strong light demander and need conditions of complete overhead light for optimum growth.

On the basis of overall observations as collected from the scattered study sites over different habitats show the nature of its distribution. The tree accommodates itself to a variety and geological formation. It thrives best on deep porous loam and although it is often found on clay. It naturally grows in a favourable manner on loose soils, derived particularly from sandstone, or mica schist.

Altitudinally this tree occurs up to an elevations of about 1000 metres or even along the valleys. It has frequently association over the slopes of hills and undulating plateau. Thus by nature of its growth the tree species is characterized by polyclimax tendency of succession.

Phytogeographical Distribution

It is deciduous plant species of tropical and subtropical forest. In the sub Himalayan tract it is a common constituent of the Salai forests and is found in fair quantity throughout the dry deciduous forest of the Indian peninsula. Sub-Himalayan tract from the eastwards, ascending to 3,000 ft., Bengal, Assam, Chhota Nagpur, central India,and the Indian Peninsula southwards to the Nilgiris,Upper Burma. As a general rule the tree is not gregarious, though often plentiful. In the sub-Himalayan tract it is a common constituent of the Sal forests, and is also plentiful on the dry waterless bhabar tract, a deep boulder formation along the base of the outer hills. Farther east, in the Duars western Assam, it is one of the commonest trees along the base of the outer hills and on the dry boulder formation skirting them, the forest being of a dry character and the chief trees besides Lagerstroemia being Shorea robusta, Terminalia Chebula, T.belerica, Dalbergia Sissoo and Acacia Catechu. In the bhabar tract of the Duars it sometimes springs up gregariously on gravel and boulder deposits near rivers, after the land has become elevated above river-level;here pure patcfhes of young Lagerstroemia, sometimes of considerable extent, may be found with large specimens of Dalbergia Sisso scattered among them, the latter being the survivors of a former riverain forest.

The tree is found in fair quantity throughout the dry mixed forests of the Indian Peninsula as far south as the Nilgiris, in association with teak, Terminalia tomentosa, Anogeissus latifolia, and other species as a rule in the Dangs forests of Surat, Large trees are occasionally a girth of 13 ft. having been recorded. In the Bori forest of the Central Provinces a girth of 15 ft. has been measured.

It is one of the common associate species of the upper storey and the over wood of both mixed and teak forests. It can grow under almost all conditions of soil and climate. It is therefore found in all vegetation climate zones of the region. The plant population distributed according to the nature of their physiographic formation. More than 70 per cent population of Lagerstromia previfolia has distributed over the slopes of the hills. In the Satpura tract it is common constituent of the forest and is also plentiful on medium to deficient rainfall, over a wide range of altitude on all aspect. Among the chief associate species of this species are Terminalia tomentosa, Adina cordifolia, Anogeissus latifolia, Diaspyros melaoxylon. In the northern part the Machna valley, Lagerstroemia parvifolia and Ouheinia delbergioides are very common species. In the Barbatpur and Betul range of Machna Valley forest tract, it is found frequently. The country is undulating with presence of a few residual hillocks. Where soil mentle is thin, poor and sandy texture thus favour the growth of this tree. The western part, the Morand and Bhangi river beds are situated on deep boulder deposits where these deposits are overlain by fertile loam the Lagerstromenia pervifolia reaches fair dimensions. In southern trap plateau it occurs in most of the forest ranges, scattered in dry deciduous forests associated with teak, Terminala tomentosa, Diospyros melanoxylon, Anogeissus letifolia and other tree. On trap soil it occurs comparatively of small size. Along the banks of nalas running through, the better qualities are found where the sub soil contains a considerable amount of sand and porous sandy loam. It is found as belt of forest either on the high ground with porous soil or along bank of the streams always as distinctly of better quality than the forest on the lower and stiffer ground away from the streams.

Table 5.4

Betul Plateau : Occurrence of Lagerstroemia parvifolia in the Mixed Forest.

Particulars	Sarni	Barbatpur	Tapti	Sawalmendha
Frequency	25.00	33.33	55.55	62.50
Relative frequency	00.34	01.50	02.13	02.21
Density	00.41	00.64	00.72	00.75
Relative density	00.68	01.23	01.90	01.75

Source: Based on field work.

Table 5.4 elucidates that the best localities of the tree are found in the southern and middle part of the region. The increasing frequency from north to south part supports the above statement. The highest frequency of occurrence is found in Sawalmendha where it has been enumerated in about 62.50 per cent of quadrats while frequency is the lowest (25.00) in the Sarani forest range. The proportion of this species in all plant species figures from 0.34 per cent to 2.21 per cent.

The individual population of the tree in a hectare is highly concentrated in Sawalmendha (0.75) and low (0.41) in Sarni but relative density is high in the Tapti (1.90) and low (0.68) in Sarni.

Applied Value

As by nature of its usefulness for the human being it is a multipurpose tree species of Central India. Nearly 37 per cent of total respondants have informed about its different uses. Negi (2003, 192-193) and Mathur (1974, 339) have also reported its uses. Its bark has medicinal properties. The wood which is extremely hard is used in the manufacturing of various articles such as plough boats etc and is also used for making agricultural implements. Its wood may be used as fuel. It also yields charcoal of very good quality.

Terminalia chebula/Harra Tree

It is middle to large sized, tree attaining 20-25 metres height with a rounded crown. Leaf but branchlets and youngest leaves are covered with soft shining and rust coloured hairs. Leaves are glabrous and pubescent, 7-18cm. by 4-8cm., 6-8 pair usually with two glands near the top. Leaf starts shedding in February-March and new leaves appear in April. Flowers, dull white in spikes at the end of branches. Fruit is more or less distinctly 5-angled, ovoid, 2-4cm. long. The plant belongs to the class of *Mesophanerophytes* and the leaf to the class of *Nanophylls*.

Natural Habitat

The tract in which this tree occurs, receives annual rainfall more than 80 cm. summers are fairly hot in this tract and the temperatures may rise to more than 40°C particularly in Satpura hills.

It is found on a variety of geological formation and on clayey as well as sandy soil. In the central India it is particularly common on metamorphic rocks. It is also noted on the Deccan trap area at base of the outer hills on deep boulder formation. It thrives well in cold plateau and moist valleys & gregariously in rather stunted form on poor rocks ground. Sandy loam soil favours its growth with sufficient moisture content, being mesophyte it grows well in moist area. Thus by nature it shows some sort of more poly climax and lesser mono-climax tendency of plant succession.

Phytogeographical Distribution

Terminalia chebula occurs almost throughout India, especially in mixed deciduous forests. Extending into forest of comparatively dry types. It ascends to considerable elevations up to 5,000 ft. in the outer Himalaya, and accourding to Bourdillon, up to 6,000 ft. in Travancore in localities where the rainfall is light. Its occurs in deciduous forests both the upper and of the lower mixed types, along with teak, Terminalia tomentosa, and their associates: It is found on a variety of geological formations, and on clayey as well as on sandy soil. Open forest or village lands,but also occurs on other geological formations . In Bombay it is common on Deccan trap, and it is noticed that on the laterite of the Mahableshwar plateau at 4,500 ft. it is one of the principal constituents of the low elfin-wood forest. It is also a characteristic tree of other special types of dry forest. The bhabar tract fringing the base of the outer hill on deep boulder formation in mixture with Sal and Lagerstroemia parviflora in a dry stunted type of forest of a pronounced deciduous character it grows gregariously in rather stunted form on poor rocky ground at about 3,500 ft. elevation, either pure or mixed with Pinus longifolia.

In the study area, it is the most important miscellaneous species of the mixed forests. The presence of the tree is scattered in nature in northern, southern and western parts of the region. It is frequently found in Landi plateau of Amla range specially in Bordehi forests, Kukru forests of Bhainsdehi range and on higher elevation of Betul range. It is also noticed in Jamun, Jhirikheda forests area of Saoligarh range and Imlikhera, Keelandeo forests and in precipitous slopes of Sarni range. It registers its occurrence in Bijadehi and Tangna forests of Sawalmendha range. In general it is scatteringly distributed throughout the mixed forests but in northern part the tree is noticed more in number. During field observation it is marked that the teak forests do not favour the growth of Harra tree. It may be concluded that it is associated with relatively colder northeastern and south-western parts of the study area.

Table 5.5

Betul Plateau : Occurrence of Terminalia chebula in the Mixed Forests

Particulars	Sarni	Barbatpur	Tapti	Sawalmendha
Frequency	41.66	35.71	27.77	37.50
Relative frequency	01.19	01.61	01.07	01.33
Density	00.47	00.50	00.33	00.62
Relative density	00.77	00.96	00.87	01.41

Source: Based on field work.

Table 5.5 makes it clear that this plant species is comparatively more frequent in the northern part rather than the middle part of the study region. The highest percentage (41.66) of frequency of occurrence is marked in Sarni range while it is lowest (27.77) in Tapti range. The relative frequency is the highest in Barbatpur (1.61) whereas the lowest (1.07) in Tapti range.

The highest concentration of density is found in Sawalmendha (0.62) range which amounts to the highest relative density (1.41). The medium density is found in Barbatpur (0.50) and in Sarni (0.47) respectively. Tapti stands at fourth place in respect of density but the lowest relative density is noted in Sarni range.

Terminalia chebula is commercially valuable plant produce, so it is collected from the forests by the forests department and exported to European countries.

Applied Values

The plant or the parts of the plant are widely used in this region as medicine 83.3 per cent of the respondents have reported its use. The brief uses and properties described by Jain (1968, 180-181), Sinha (1996, 86-87) and Saxena (1970, 123) are as follows. Fruit is dry and heating, stomachic, anthelmintic alterative and useful in asthma, sore throat, eye diseases and disease of the heart and the bladder. The bark has diuretic and cardi tonic properties. It is usually used in combination with Aonla and Bahera which is known as 'Triphala'. The dried fruits of the tree constitute the drug commonly known as Chebulic Myrobalan. This can be applied externally on chronic ulcers, wounds and scalds or used as a gargle in inflammation of mucous-membrane of mouth. It has also been found to have some effect on blood pressure as cardiac tonic. The powder of the fruit is used as a dentifrice for the strength of gums. Wood is very hard, fairly durable, used for making agricultural implements, crates fence posts and other purposes. The tree is important mainly on account of its fruits, which are the best of the commercial Myrabolans used for tanning.

Madhuca indica/Mahua Tree

It is a large deciduous with well developed spreading crown and canopy is of oval shape. Trees may attain height of more than 20 metres. Bark is thick, wrinkled and of grey to brown coloured, cracked. The inner bark is red milky. The trunk is short with numerous spreading braches forming a thick shady head. Leaves are clustered near the ends of branches, hard and densely pubescent. 12-20 cm long. Young leaves appear in spring and are of copper colour. Flowers cream coloured grow in dense fascicles near the ends of the branches below the terminal leaf bud. Flowers bloom from February to April when the tree is almost leafless. The fruits ripens during June to August and fall by itself. From leaf-classes point of view it falls under the class of *Mesophyll* while under the classification of life-forms the plant comes in the group of *Mesophanerophytes*.

Natural Habitat

The climatic conditions favour the growth of this species vary from tropical to sub tropical. In the natural habitat the absolute maximum shade temperature varies from 40°C to 12°C and normal rainfall is 75cm. to 175cm. It is not found in the dry regions. It grows in all type of soil, resulting from the decomposition of various rock formation but in friable loam soil its occurrence is noted much more where these trees put an optimum growth. It grows wild in forests as well as in cultivated areas, of villages in common and on other type of wasteland. Sandy plains are one of the most favourable habitats for its growth and development. Thus by nature of plant growth the traitor plant species is characterized by poly-climax tendency of plant succession.

Phytogeographical Distribution

This is a very common plant throughout Central India and sub-mountain Himalayan region.

Madhuca indica is distributed throughout the study area. It thrives well on both the level and the lower slopes. As it is one of the associate species of the mixed forests, its presence is more frequently marked in mixed forests rather than the teak forests. It is abundant in Kanjitalao felling series with predominantly mixed type of forests in Barbatpur range. In Sarni range, deep soil areas like Salaiya and Loniya village and waste land of Dodramuhal in Tawa valley show its occurrence as pure crop. In Sawalmendha range, it is found in Pathakheda, Ramghati and Khumai forest villages. It is also marked along roadside from Ghodadongri to Sarni. It is ordinarily drought and frost hardy but suffers under severe conditions. Now-a-days it is also preferred for social forestry plantations in study area.

It is evident from table 5.6 that this tree is one of the most common trees in the study region. Its presence is recorded in more than two fifth of the sample quadrats. Frequency of occurrence is high in southern parts of the region where it is 43.75 per cent, in Sawalmendha range followed by Barbatpur (42.85). Sarni and Tapti range show similar frequency which is low (38.88). The relative frequency 1.94 is the highest figure of Barbatpur.

Table 5.6

Betul Plateau : Occurrence of Madhuca indica in the mixed Forests

Particulars	**Sarni**	**Barbatpur**	**Tapti**	**Sawalmendha**
Frequency	38.88	42.85	38.88	43.75
Relative frequency	00.89	01.94	01.49	01.55
Density	00.50	00.57	00.44	00.56
Relative density	00.82	01.10	01.16	01.27

Source: Based on field work.

The highest figure of density 0.57 is for Barbatpur while the lowest (0.44) is Tapti range. The medium density is found in the Sawalmendha and Sarni range. The relative density is as high as 1.27 in Sawalmendha and as low as 0.82 in Sarni range. This discussion can be concluded that Mahua tree is more frequent in the northern and southern parts of the study area. In central part the growth is stunty.

During survey it is marked that eastern wasteland of Dodramuhal railway station between Itarsi-Betul track, Madhuca indica counts 107 per hectares. In this way it has been seen that in some parts of the region, it grows as dominant species.

Applied Values

The species ranks fifth in order of its meritorious use. According to Chopra (1956, 159), Dastur (1964, 109-110), Saxena (1989, 215) and wealth of India (1962) the different uses and properties of it are described here. The bark of tree is reported to heal wounds and ulcers, cures leprosy and used in fractures. The milky juice from the bark is astringent, hastens suppuration and cures disease of the blood. The flowers yield a distilled spirit which is astringent, tonic and appetizing. Mahua oil has emollient properties and is used in skin disease, rheumatism and headache. It is laxative and considered useful and habitual constipation, piles and haemohoidea. It is also used as an emetic *(Suri and Mathur 1988)*. The seed cake is also used in snake bite.

Flowers and fruits are nutritive and edible by local residents. They may also be fermented and distilled for spirit. Fragrant flowers attract honeybees. Pulp is used for extraction of oil. Wood is moderately hard and used in the manufacture of plywood and paking cases. The wood is used as a fuel too. It is ornamental, avenue and short tree extremely suited for forestation of degraded area and wasteland.

Albizzia lebback/Siris Tree

It is large unarmed deciduous tree with a well-developed crown and straight, clear bole. It attains a height of 20 to 36 metres. The bark is greyish with many short irregular cracks. The heart wood is dark brown, tough and durable. The old leaves fall in winter and new leaves appear of the end of the winter season. Flowers are white and arranged in globose umbellate heads, appear chiefly in April to May. The pods ripen develop in August some are nearly full sized. The plant belongs to the class of *Mesophanerophytes* and the leaf to the class of *Monophylls*.

Natural Habitat

The tract in which this tree occurs experiences very hot summer with the mercury soaring to more than 40°C. The total annual rainfall is more than 70cms. For its best development the tree requires full over headlight. The presence of Siris tree is registered mainly on flat surface and river belts along the lower slopes of hills and foothills, up to an elevation of about 15 metres. It is not exacting as to soil and will grow fairly well even on laterite or black cotton soil with sufficient moisture content. It is usually occurred scattered and not gregariously.

Phytogeographical Distribution

The tropical and subtropical forests of India are characterized by the presence of this tree. It grows mostly in deciduous forests of the study area. The tree is more commonly found in the central part rather than the northern part. It is uniformly noticed on the Nishan river tract especially in Nanda forest circle of Chicholi range. The growth of tree is unevenly marked in Ladia-aam of northern Sarni compartment and Kalwadi in Barbatpur range. In the mixed forests of low lying areas, it is very scarcely registered. Usually it is a planted species too. It is cultivated in Bagdona village along the roadsides on Ghodadongri-Sarni road.

The data appended in the table 5.7 elucidates that the occurrences has been noted in half of the quadrats in Tapti range, therefore, frequency (50 per cent) is the highest in this range. It is less than one third in Barbatpur range (28.57) but in Sarni the frequency figure is lowest. In respect of the relative frequency Tapti is at the top, medium in Barbatpur whereas Sarni again is at bottom. The species is not marked in the Sawalmendha range.

Table 5.7

Betul Plateau: Occurrences of Albizzia lebback in the Mixed Forests

Particulars	Sarni	Barbatpur	Tapti	Sawalmendha
Frequency	13.88	28.57	50.00	00.00
Relative frequency	00.40	01.29	01.92	00.00
Density	00.19	00.28	00.55	00.00
Relative density	00.31	00.54	01.45	00.00

Source: Based on field work.

The density per quadrat is highly concentrated in Tapti (90.55) and the lowest (0.19) in Sarni. Accordingly the relative density is also found.

The survey conducted has shown that in the central part, most of the survey sites registered presence of Siris tree. In the study area, the tree is likely to disappear because the local people cut the wood in order to make cart-wheel and agricultural implements.

Applied Values

The Albizzia lebback is being used by 21.49 per cent of the persons interviewed. Chopra and others (1956, 11), Gupta (1990, 111) and Mathur (1956, 101) has mentioned the medicinal uses of the plant which are summarized here. The bark of the tree is anathelmintic and relieves toothache, strengthen the gums and the teeth. It also cures diseases of the blood. The bark and seeds are astringent too and given in piles and diarrhoea. The flowers are applied externally in boils eruptions and swellings. The leaves are given in night blindness.

The timber is strong hard and durable. It may be used for the interior decoration of buildings, flooring and paneling. Other uses of its wood are for making veneers, plywood, boats, musical instruments, mine pit props and agricultural implements. The leaves and twigs of the tree are an excellent folder.

Terminalia bellerica/Bahera Tree

A large tree with straight bole and a well developed crown. It attains a height of more than 25 metres. The bark is bluish or brownish with characteristic long narrow cracks. Leaves gathered about the extremities of the branches and are glabrous; usually punctuate on the upper side, 10-22cm. long which appear in April. Flowers are greenish white appear in April to May with young leaves. Fruit is ovoid, grey, velvety, 2-3cm. long densely covered with hairs. From life forms point of view the tree falls under the group of *Mesophanerophytes* but from leaf classes point of view it falls under the class of *Microphylls*.

Natural Habitat

It grows in wide range from tropical to subtropical region. In its natural habitat the temperature varies from 36°C to 46°C and the temperature remains low in the winter season. More than 80cm. rainfall is received in the tract in which Terminalia bellerica grows. It is very common in mixed forests of flat plateau areas and plain areas. Though it grows in all kind of soils, but in soil derived from metamorphic rocks the growth seems to be good. In the Indian Peninsula it occurs most frequently in moist valleys and extending to the adjoining tracts. Best development of this tree is observed in the precipitous slopes.

Phytogeographical Distribution

This tree occurs almost throughout India, excepting the dry regions of western India. It is found in the sub Himalayan region, mainly along the transition zone with the Indo-Ganga plains and Central India in the mixed forest zone.

Among several miscellaneous species of Betul forests the Bahera tree is the most important one. It is found in mixed forests of all the forests ranges, although grows well in moist type miscellaneous forests of the area under study. Generally, it is scatteredly distributed in the region but in the northern parts the tree is often marked in compact patches particularly in Bordehi forests of Amla range, Tara and Menda Khera forests of Barbatpur range. In southern part it is noticed in Junapani forests of Sawalmendha range. Bahera tree is scarcely found in mixed forests of lower humid slopes to the west particularly in Nanda forests of Chicholi range, Batki forests of Mohda range, Pakshi and Pat forests of Bhaura range and precipitous slopes of Landi.

Table 5.8

Betul Plateau : Occurrences of Terminalia bellerica in the Mixed Forests

Particulars	Sarni	Barbatpur	Tapti	Sawalmendha
Frequency	19.44	21.42	11.11	18.75
Relative frequency	00.56	00.96	00.43	00.66
Density	00.19	00.36	00.16	00.31
Relative density	00.31	00.69	00.42	00.70

Source: Based on field work.

Table 5.8 shows that the frequency of occurrence of Bahera tree is almost the same throughout the mixed forests of the study area. Even then frequency of occurrence is high (21.42) in Barbatpur range followed by Sarni and Sawalmendha range i.e. 19.44 and 18.75 respectively. The lowest figure (11.11) is in Tapti range. Similarly the relative frequency is high in Barbatpur and low in Tapti range.

Proportion of the plant in total plants occurring on sample sites is relatively low. The density is the high in Barbatpur (0.36) range but the relative density is highest (0.70) in Sawalmendha range. The lowest density is noticed in Tapti range (0.16) while the lowest relative density (0.31) is in Sarni Range.

On the basis of above discussion it can be concluded that this species is though common but scanty. Both, Terminalia bellerica and Terminalia chebula grow in the same ecology even though the number of Terminalia bellerica is rather less than Terminalia chebula in the study area.

Applied Values

It is seen that Terminalia bellerica stands fourth in order of its utility for different purpose Kirtikar and Basu (1933, 1017-1020), Singh and Khan (1990, 138 & 155) and Oommachan (1991, 203) have enumerated the medicinal characteristic of the Terminalia bellerica in their studies. Important among them are as follows. The dried fruits are used in stomach disorder, bronchitis, sore throat, inflammation and in disease of the eye, nose, heart and the bladder. It is also given as a brain tonic and applied to piles. The seeds are useful in thirst corneal ulcers and relieve vata. The oil is considered a good applicant for the hair. The gum is believed to be demulcent and purgative. Terminalia bellerica is also useful in leprosy, dropsy and fever. The bark is used to relieve pain in chest muscles and congestion by cough. During field survey it is marked that the local people use its oil as an applicant for rheumatism.

The wood is strong and durable. It is used in buildings construction, and boat making.

Diosyros melanoxylon/ Tendu Tree

It in middle sized deciduous tree reaching 15 metres in height. It is conical crown with spreading branches. The bark is dark grey or black, exfoliating in rectangular scales. Leaves are oval, 8-14 by 4-7, from leaf-classes point of view the plant falls under class of *Microphylls*. Flowers are short, yellowish white, flowering season is February to April. Fruits are yellow to orange coloured when ripening, ovoid or globose. From life forms point of view the tree falls under class *Mesophanerophytes*.

Natural Habitat

In climatic requirements it shows a wide range. In its natural habitat the maximum temperature is 38°C to 45°C and the minimum ranges from 6°C to 15°C. The tract in which this tree occurs, receives annual rainfall more than 70cms. It grows under all kinds of soil. In poor soil areas its growth is registered well. In the Central India it is one of the common species in a poor stunted type of forests where the soil has an excess of calcareous with reddish clay on the surface.

On dry foot hills it is found in places where the soil is shallow or gravelly. It is, therefore, found in all plant environments.

Phytogeographical Distribution

The tree is common throughout the greater part of the Indian Peninsula, in dry type of mixed forest on a variety of geological formation and soil. Specially in Central India it is frequently found mixed with a variety of deciduous species not only in mixed forest but also as an association with teak and sal.

The tree is seen all over the study area. Its occurrence is noticed well in dry deciduous forests and marginal land of forest. It shows its presence frequently in sandy soil areas such as Asir hills, and Morand plateau toward the north. It is found in colonial forms in the Rampur Bhatodi project on yellowish graval soil areas. Its occurrence is also observed in shallow soil areas toward the southern part and in the eroded areas of Tapti Valley. It is common on undulating land of the southern part i.e. Bhainsdehi plateau. A good stand of it is registered on western margin of high Chicholi plateau. It is also noticed that the occurrence of the tree is more common in dry teak forest area of southern part. It is observed that the density and condition of plant growth varies from one place to the other, due to their occurrence at different habitat characteristics.

Table 5.9

Betul Plateau : Occurrences of Diosyros melanoxylon in the Mixed Forests

Particulars	Sarni	Barbatpur	Tapti	Sawalmendha
Frequency	22.22	21.42	66.66	50.00
Relative frequency	00.63	00.97	02.56	01.77
Density	00.38	00.57	01.16	01.06
Relative density	00.62	01.09	03.07	02.41

Source: Based on field work.

Table 5.9 indicates that the tree occurs in at least one fifth (21.42 per cent) of the quadrats surveyed in Barbatpur range and as high as (66.66 per cent) quadrat in Tapti range. Proportion of this species in all plants is known as relative frequency ranges from only 0.63 per cent in Sarni to 2.56 per cent in Tapti. Thus it is more frequent in central and southern part of the study region.

Applied Values

Nearly one-third (about 29.8 per cent) of the total respondants interviewed have informed the uses of the plant.

Diosyros melanoxylon leaves are used for bidi making industry therefore, its leaves are collected by forest department every year. It is also used as a fuel, particularly in the rural areas. The ripen fruits of the Diosyros melanoxylon are very sweet and widely eaten locally. Ripen fruits cure diseases of blood. Dastur has (1962, 74) given the medicinal account of Diosyros melanoxylon. The bark is good for dysentry and the juice for the fresh wounds. The seeds are preserved by the country people and given as an astringent in diarrhoea.

This species is likely to disappear because the Tendu patta collectors not only pluck the leaves but also cut the branches in order to get the leaves.

Butea monosperma/Palas Tree

Tree may go as high as 8-16 metres, main trunk is strong, crooked with irregular branches. The bark is rough and ash to grey coloured. Leaves of 3 leaflets, broadly ovate, hard, rough surface on the upper side while Tomentose on the lower side. The leaves of the tree fall off in winter and appear in April to May. The tree and its leaf belong to the class of *Mesophylls*. Its scarlet and orange flowers come in such profusion that the tree has been very aptly named as the flame of the forest, flowers appear in about February to March in small but dense clusters generally on leafless branches, during this period the tree appear to be loaded with a mass of orange colour. The fruit is a flat pod having a single seed.

Natural Habitat

Eco climatically this tree is distributed in a very wide range, including nearly all type of habitats. Due to its drought resistant character its growth as well as development coincides except in semi arid habitats. The rock and soil on which the tree grows have a marked influence on its growth. The tree is naturalized in heavy soil. The large dimensions in which it reaches are trap formations, black cotton soil though stiff soil, in forming pure crop of greater or less extent where the trees which are frequently in the pole or sapling stage may grow thickly together. It is more common in plains or ravines and water logging areas. It also thrives in agricultural waste land and marginal land.

Phytogeographical Distribution

This tree has very wide distribution over Indian sub continent and is chiefly distributed in the mixed or dry deciduous forest of central and western India.

It is a very common and well known tree in the study area. It is one of the usual associate of miscellaneous forests and is found throughout the forest tracts of the region. Butea monosperma is frequently seen in the central part in comparison to northern and southern parts because of black soil cover. The growth of this species has been recognized as a subtype in the forests which is registered in the Tapti range from Kheri to Tapti River. It is found as a dominant species in deep ravines of Madamjhiri in Betul range. It is also noticed in the mixed forests of Bhaura range as associate species especially on lower slopes of northern escarpment. A greater part of the Betul plateau is associated with the black cotton soil which favours the growth of Butea monosperma. The good crop is seen almost in agricultural waste land of the study area. The tree is also marked along the roadsides all over the region.

Table 5.10

Betul Plateau : Occurrence of Butea monosperma in the Mixed Forests

Particulars	Sarni	Barbatpur	Tapti	Sawalmendha
Frequency	38.88	57.14	66.66	56.25
Relative frequency	00.89	02.58	02.56	01.99
Density	00.44	00.71	00.72	00.69
Relative density	00.72	01.36	01.89	01.57

Source: Based on field work.

The data appended in table 5.10 elucidates that the occurrence of the Palas tree has been noted in two third quadrats in Tapti range. Therefore, frequency (66.66) is the highest in this range. It is more than half surveyed quadrats in Barbatpur (57.14) and Sawalmedha (56.25) but the one third quadarts in Sarni range in respect of the relative frequency Barbatpur is at the top and Tapti stands second with 2.56 per cent. Sawalmedha has medium relative frequency and Sarni comes to the bottom with 0.89 per cent.

The density per quadrat is more numerous in Tapti (0.72) while it is least in Sarni (0.44). Accordingly the highest relative density is in Tapti (1.89) and the lowest in Sarni (0.72) forest range. The medium relative density is marked in Sawalmedha (1.57) and Barbatpur (1.36).

During the course of survey it has been noticed that the tree is highly concentrated in the central part of the survey area. The frequency of the tree gradually decreases toward the north and south which indicates that the hill do not favour the growth of Butea monosperma.

Applied Values

It is the most useful as well as a multipurpose species. About 26 respondents out of 84 have informed the different use of this weed. Dastur (1964, 41) Khan, et.al. (1988) Bentley and Trimen (1981, 79) and Parikh (1999, 243-263) have also stated the use and medicinal properties of the plant which are as follows. It is used in several combinations both as an anti-helminthic and as an anti-fertility drug. It is used for eradication of intestinal worms and improves the function of stomach and intestine. The bark of the tree is useful in fractures of the bones and disease of the anus. The leaves are good for diseases of the eye. The gum contain tannins and is valuable for treatment of diarrhoea. The seeds are largely used in the treatment of roundworm. A secretion of the tree "Kamarkas Gond" is considered as very useful substances for the treatment of female diseases. It is a boon for a woman who fails to conceive. It gives strength to the female urino-genital and sex organs and particularly to the womb (uterus).

The tree has economic importance because a large number of useful products such as gum, lac, dye etc. are obtained from it. The local people (rural areas) collect its leave for making Pattals and Donas (leafy vessel used to serve dishes).

It is host plant of Lac insects. They draw nutrients from the sap. The roots of this species yield a fibre that may be used for making ropes by the inhabitants. They also uprooted the plant for its root, as they make ropes and brushes from it. In this way the plant is gradually disappearing from the forests of the study area. So there is a need to protect the species from disappearance.

Terminalia arjuna/Arjuna Tree

This is a evergreen or semi evergreen tree with a large crown and dropping branchlets. It attains a height of more than 25 metres. Bark smooth, exfoliating in thin irregular sheets, green when nearly exposed, turning light grey, pink inside. Leaves are oblong 10-15 cm. long, narrow leathery, petiolate. Thus, due to its leaf size, plant belongs to class of *Microphylls*. Flowers are white, appear from April to July and fruits ripen following February to May. The plant belongs to the class of *Mesophenerophytes* but some time it belongs to class of *Megapheneropyhtes* from life forms point of view.

Natural Habitat

It in found naturally in regions where the maximum temperature varies from 38°C to 48°C, while winters are mild. The total annual precipitation is more than 80cms. however, its occurrence depends largely on the moisture supplied by streams. Its distribution is not governed by climatic consideration alone.

It is characterized by *Mesophytic* habits due to which habits it is frequently found on the banks and slopes of water courses like rivers, streams, ravines and in the inner dry valleys beyond the reach of monsoon rains. This tree also noticed on humid slopes and low lying areas with sufficient moisture content. It reaches a large size in valleys on fertile alluvial ground and becoming gregarious on alluvial loamy soil. It shows a marked tendency to occupy the more level ground where the soil is deep and moist. Thus, by nature we can say that the tree shows more monoclimax tendency of plant succession.

Phytogeographical Distribution

Terminalia arjuna occurs almost throughout India chiefly along swampy areas. It is common in Central India, Deccan plateau and sub Himalayan tract.

It is registered in most of the forest ranges as well as water logging areas of the study region. It occurs in scattered patches, especially in areas along river and nalas and in areas with greater soil depth. In northern part it is found near the lower banks of Machna River especially in Banka and Kantawadi villages. In the Northeastern and northwestern part along the river Tawa and Machna respectively, the growth of Kahu tree is marked very well. The central parts drained by Tapti and its tributaries, also show its occurrence. Though its occurrence is noted throughout the study area but it is commonly and uniformly found in mixed forests of moist-type. In early Sanskrit works, this tree has been named "Nadisarjja", which denotes that the tree grows on banks of river. It is unevenly found all over the region.

Table 5.11

Betul Plateau : Occurrence of Terminalia arjuna in the Mixed Forests

Particulars	Sarni	Barbatpur	Tapti	Sawalmendha
Frequency	19.44	35.71	33.33	12.50
Relative frequency	00.56	01.61	01.28	00.44
Density	00.28	00.43	00.50	00.19
Relative density	00.46	00.83	01.31	00.43

Source: Based on field work.

The data summarized in Table 5.11 shows that Kahu is highly concentrated in the northern part of the study region. The highest percentage of occurrence (35.71) in the quadrats surveyed is found in Barbatpur followed by Tapti range (33.33). The lowest frequency is found in the Sawlmendha range. The relative frequency also follows the same pattern.

The tree is unevenly distributed in the study region. Due to its natural habitat it grows only in a particular type of habitat or ecology.

Applied Values

About 60.7 per cent respondents have informed that they are prescribing the drug made out of Kahu tree. Kirtikar and Basu (1933, 1023-1028) Ambasta and others (1992, 627) and Parikh (1986, 88-89) have also described the uses and drug properties of the species. Its bark is alexiteric tonic and antidysentric useful in fracture, ulcers, blood diseases, urinary discharges and diseases of the heart. The bark is astringent and febriguge too. Pulverized bark gives relief in symptomatic hypertension and act as a diuretic in cirrhosis of liver. The fruit is tonic and deobstruent. The juice of the fresh leaves is a remedy for ear ache. Ghoshal (1909) was the first to investigate the drug in heart diseases. Arjunarishta is the most famous tonic in Ayurvedic.

The wood of tree is strong and durable. It is suited for building construction, industrial timber and for making agricultural implements. It is also used for making boats. The wood is used as a fuel. The leaves are evergreen and the tree affords a good fodder for goats.

Other Valuable Species

Besides the above described species some other significant species are also found in the mixed forests of the area under study. Some of them are Syzyium cumini (Jamun), Schleichera oleosa (Kusum), Salmalia malabarica (Semal), Mangifera indica (Aam) trees, Nyctanthus arbortristis (Harsingar), Helicteres isosa (Marorpali), shrubs, Tinospora cordifolia (Gilloy), Gymnema sylvestre (Gurmar) climbers and Asparagus adscendens (Musli) herb are well in number near alluvial plains and deep soil areas with greater moisture retentivity while the Semecarpus anacardium (Bhilwa),

Table 5.12

Betul plateau : Other valuable plant species in Mixed Forests

S.N.	Species	Particulars	Sarni	Barbatpur	Tapti	Sawalmendha
1.	Semecarpus anacardium (Bhilwa) 'T'	Frequency	27.77	21.42	16.66	43.75
		Density	00.33	00.28	00.16	00.56
2.	Bauhinia variegata (Kachnar) 'T'	Frequency	36.11	35.71	27.77	12.50
		Density	00.47	00.43	00.33	00.25
3.	Syzygium cumini (Jamun) 'T'	Frequency	33.33	21.42	22.22	06.25
		Density	00.41	00.36	00.27	00.12
4.	Schleichera oleosa (Kusum) 'T'	Frequency	25.00	14.28	50.00	62.50
		Density	00.28	00.36	00.66	00.81
5.	Nyctanthus arbortristis (Harsingar) 'S'	Frequency	38.88	21.42	11.11	18.78
		Density	00.50	00.50	00.27	00.37
6.	Helicteres isora (Mororphali) 'S'	Frequency	22.22	35.71	27.77	31.25
		Density	00.33	00.64	00.50	00.44
7.	Tinosopra cordifolia (Gilloy) 'C'	Frequency	25.00	21.42	16.66	12.50
		Density				
8.	Asparagus adscendens (Musli) 'H'	Frequency	00.00	21.40	00.00	12.50
		Density				

Source: Based on field work.

F = Frequency D = Density T = Tree

S = Shrub H = Herb C = Climber

Bauhinia variegata (Kachnar), Aegle marmelos (Bel) trees, Aloe vera (Ghritkumari) shrub, Celastrus paniculata (Malkangni) climber are found in poor and shallow soil area on hilly tracts.

Some species of greater commercial value can also be seen in the area under study. Musli is a very significant and popular herb, Pandey (1991, 63) and Bajaj and Mishra (1990, 159) have elaborated the uses and properties of this herb. The roots are used as a general health tonic for women given to them after delivery. Gilloy, Gurmar leaves and Jamun seeds is very good medicine for diabetes. Sinha (1996, 89), Bantly and Trimen (1981, 12) have reported them.

ENVIRONMENTAL IMPACT ASSESSMENT

To know the distribution of different species in mixed forests of the study area, it is necessary to study the ecology of these forests, as ecology helps to investigate the environmental adaptations, species regeneration and economic importance too. Mixed forests consist of mostly middle aged crop with mature trees scattered in the forests.

It is possible to find mixed forest of varied quality, classes and density at various places, as a result of edaphic factor. In moist mixed forests luxuriant growth of different life forms are found densely. The site quality being from II to IV. The density ranges from 0.6 to 0.9 with average density being 0.7, while in the dry mixed forest density ranges from 0.5 to 0.7 with site quality varying from III to IV. Life form spectrum of the plant species for these forests shows the Phanerophytic nature.

The forests of the river valleys with alluvial soil are being cut for the settlement and cultivation. The forest areas being surrounded by populated villages are subjected to heavy pressure of pasture and grazing. Thus, those forest which are in the vicinity of settlement are poor due to excessive cutting, over-grazing and repeated fires.

These forests have been repeatedly cut and over exploited to meet the increasing demand of fuel wood and various minor forest products by the forest department from a long period. The regeneration of trees is not satisfactory, therefore, the crop has degraded and is uneven aged, the bulk being middle aged with a few scattered old trees. In this way the ecology of mixed forests is leading to the "Seral" stage. It has, however, been observed that the ecological status of the mixed forests probably appears to be a biotic climax.

CONCLUSION

The areas where the percentage of miscellaneous species is about more than 16.3 per cent of the total forested area, ecologically termed as mixed forests. In the study area these forests are found in three belts. The maximum proportion of these forests is marked over Gawaligarh hills in the south division. In the north division, they spread over Asir range and northern escarpment towards the east-west. In these forests different layers are formed with various kinds of associates.

The study area has a rich and varied flora on account of its diversified topography and related to other variable environmental factors. The variation in topography factors change the floristic composition and plant diversity in the region. Important areas, tremendously rich in various types of medicinal trees are the Asir range, Tawa Basin, Central High Plateau, Tapti Valley and Gawaligarh hills.

The areas which consists soil derived from metamorphic crystalline rocks possess vast species diversity and it is a good sign for occurrence of numerous plants. The forest layer formations with alluvial soil having more number of species.

The distribution of trees at different sites of study area shows that the species are not evenly distributed due to micro climatic change. The moist localities of the region are occupied by the most valuable species such as Terminalia chebula (Harra), Emblica officinalis (Aonla), Terminalia bellerica (Bahera), etc. the number of these species is good in Landi and Khamla plateau whereas comparatively drier parts i.e. the southern scarpment of the study area have been occupied by Diospyros melanoxylon (Tendu), Lagerstromia parviflora (Lendia), Cassia fistula (Amaltas) etc.

Certain mesophytic trees of semi-evergreen nature such as Terminalia arjuna (Kahu), Bauhinia variegata(Kachnar),Ficus glomerata (Gular), Scheichera oleosa (Kusum) etc. are registered in well drained areas and are confined to the banks of river and nalas. These localities can be remarked for the growth of plants. Some unique plant species are widely distributed in the forests as herbal remedies of various diseases. These valuable medicinal plants grow extensively in the mixed forests of Betul plateau. Trees such as Siris, Palas and Amaltas are in an alarming stage in the forests of study area. These species are being cut by the local people and traders for fuel wood. Some species are declining due to lack of natural regeneration including Emblica officinalis (Aonla),Terminalia chebula (Harra), Madhuca latifolia (Mahua), Scheichera oleosa (Kusum) etc. Emblica officinalis and Terminalia chebula and Diospyros melanoxylon have suffered because of fruit collection by local people and forest department, on the other hand Madhuca latifolia flowers are known for use in preparation of distilled liquors and its seed is used to extract oil. In this way breeding of various species have been continuously on decrease. This over exploitation and mis-use of the resources has resulted into extinction of these important plants from habitat. So it is urgently needed to undertake phytogeographical studies of rare and vanishing species in natural forests in order to protect them in their natural habitat.

REFERENCES

Ambasta, S.P. and Others, 1992, *The useful Plants of India,* New Delhi, Publications and Information Directorate Council of Scientific and Industrial Research.

Bently, Robert and Triman, Henry, 1981, *Medicinal Plants*, Vol. II 78, and vol. III 174, Dehradun, Allied Book Centre.

Chopra, R.N., Nayar, S.L. and Chopra I.C., 1956, *Glossary of Indian Medicinal Plants*, New Delhi, Council of Scientific and Industrial Research.

Dastur, J.F., 1962, *Medicinal Plants of India and Pakistan*, Bombay, D.B. Taraporevala Sons and Co. Ltd.

Gupta, S.P., 1990, Native Medicinal Uses of Plants by the Asurs of Netarhat Plateau, in *Contribution to Ethnobotany of India*, ed. By S.K. Jain, Jodhpur, Scientific Publishers, 101-114.

Jain, S.K., 1968, *Medicinal Plants*, New Delhi, National Book Trust of India.

Khan, M.A., et. Al., 1988, Some Herbal Drug used in Helminthic Infestations, Paper Present to & National Symp. On the Development of Indigenous Drugs in India, New Delhi, April 8-10.

Kirtiker, K.R. and Basu, B.P., 1933, *Indian Medicinal Plants*, Dehradun International Book Distributors.

Mathur, R.C. 1970, *Systematic Botany*, Agra, Agra Book Store.

Oommachan, M. and Masih, S.K., 1991, Conservation and Cultivation of Excessively Exploited and Vanishing Medicinal Plants of Madhya Pradesh, Focal them ISCA Symposium , Delhi, Narendra Publishing House, 197 – 212.

Negi, S.S., 2003, *Natural Resource Management in the Himalaya*, Vol. II., New Delhi, A.P.H. Publishing Corporation.

Pandey, R.K. and Shrivastava, J.L., 1991, Wild Medicinal Plants of Fabaceae in Madhya Pradesh, An Ethnobotanical Approach, *Journal of Tropical Forestry*, July-Sept., Vol. 7(III), 221 – 225.

Parikh, Gopinath Vaidya, 1999, Neem, Van-Aushdhi Ratnakar, Vijyaygarh (U.P.) *Sudhanidhi Karyalaya*, Feb.-March, Vol. 5, 162-196.

Parikh, Gopinath Vaidya, 1986, *Van-aushdhi Ratnakar*, Vijyaygarh (U.P.) Dhanwantari Karyalaya, Vol. i.

Saxena, K.P., 1970, *A Brief Course in Economic Botany,* Lucknow, Prakashan Kendra.

Saxena, R., 1989, Use of Mahua Syrup as a Source of Sweetening Agent in Snack Preparation of Tribals, in *Tribal Ecosystem and Malnutrition in India*, ed. By P.D. Tiwari and A.N. Sharma, New Delhi, Northern Book Center, 215-220.

Sharma, R.N., 1989, *Dehati Jadi Butian*, Delhi, Hindi Pustak Bhandar.

Singh, V.K. and Abraham Khan, 1990, *Medicinal Plants and Folklores* (A Strategy Towards Conquest of Human Ailments), New Delhi, Today and Tomorrow's Printers and Publishers.

Sinha, Rajiv K., 1996, *Ethnobotany*, The Renaissance of Traditional Herbal Medicine, Jaipur, Ina Shree Publishers.

Wealth of India, 1962, New Delhi, Raw Materials Council of Scientific and Industrial Research. Vol. VI.

Sharma, A.K., 1988, Biotic Resources of the Ranthambhore National Park in *Environmental Resources*, Add by H.S. Mathur, Jaipur, RBSA publishers, 91-102.

6

Ecological Status of Salai Forests

Nature has endowed Satpura with varied natural vegetation. The Satpura relief and the total landscape determine the distribution of plant communities and provide habitat. The present communication reports are observations on the standing crop structure and composition, succession trend and above-ground biomass accumulation in woody species growing at rocky slopes of hills, ridges, and dry top of hills in the study area. In these habitats, generally soils are found thin, largely deficient in minerals and nutrients due to constant washing by the rains. Soil texture is sandy to sandy loam which has been occupied mainly by Salai forests. Thus forest crop tends to be open and of poor quality as savanna type. Soil decomposition, release of nutrients and leaching is a very slow process and so improvement in site quality and density of stocking of species in slow. Blatter and Halbery (1918, 21) termed such vegetation type as "formation" which are exclusively controlled by edaphic factors.

In the preceding chapter an assessment of the Salai forests resource of Betul plateau has been made. Benefits derived from the Salai forests are many since species is multipurpose in nature. They protect us from various calamities and provide us many raw material. According to Guthric (1961, 51) logs are the basic raw material for industries which convert them into products such as lumber, pulp and paper. For example Nepa Paper Mills Nepanagar gets sufficient raw material from the Salai forests. Salai wood is also used for packing wooden boxes handles of agricultural tools etc.

Salai forests have repository of plants having medicinal value which have been used by human beings. Local communities are often using these plants for their basic needs as well as for commercial purpose even today. Rusa oil is extracted from the Saphia and Motia grasses occur in the Salai forests which is useful for soap, cosmetics and pharmaceutical industries.

The principal forest products obtained from Salai forests include timber, small wood, fodder, fuel wood, fruits and many other products of lesser value. These forests are usually open with tree growth interperred by grassland that are used as grazing land in rainy reason. During monsoon, grasses grow rapidly in these forests hence the forest dwellers (Gouli) stay here along with their cattles in camps. More than 30 per cent of the total population depends on fuel wood or charcoal for their energy needs which they collect from forests. Tribals and inhabitants of villages near forests collect fruits like Buchanania lanzan (Achar), Aegle marmelos(Bel), Emblica officinalis (Amla) etc. from the forests for their use as well as for selling them to earn money.

Forests maintain equillibrium between the useful and harmful gases in nature. The healthy effects of forests on human life are of greater value as they purify our natural environment. *(Singh, 1987, 51)* More space for forests will be able to neutralize the hot house affected concept. Betul forests also protect the land from solar radiations to a great extent which results in lowering temperature. In Betul plateau modification of micro climate is affected by forests cover. Its micro-climate indicates that rainy season is more equitable while summer season is comparatively milder than the surrounding regions.

ECOLOGICAL REQUIREMENTS

Eco Climate

Growth of Bosewellia serrata (Salai) by seed and coppice are highly susceptible to high summer temperature with low rainfall. This type of forest colonizes in areas having deficient rainfall. The tract in which these forests occur experiences very hot summers with the temperature being more than 43°C. Micro climate of plateau varies place to place according to attitude, aspect and configuration of the ground. The temperature is more equitable in plateaus, than in valleys, the days are hotter in southern aspects. Monsoon winds do not get easy access across the Satpura range, resulting in comparatively low rainfall in southern plateau. It is due to these favourable factors the natural growth of Salai forests is in plenty in the southern plateau. The precipitation and evapotranspiration determine the soil moisture content which is also shown graphically in figure (Plate 6.1).

From mid June to mid December i.e. during a period of six months the soil moisture conditions is optimum to fair. Climate is hot to warm and desicating to damp, while the general conditions being favourable for plant growth.

Natural Habitat

Salai forest occurs in physiographical dry soil and shallow soil on all geological formation of which the chief are sandstone, Shale, Gneisses, Mica schist and laterite. Sandstones disintegrate into soils of low fertility, except in areas where weathered material from intruded or adjacent igneous or metamorphic rocks have added some fertility on which Salai can thrive. The Central India comprises the most part of its forests some what dry type, and hence the occurrence of Salai is determined largely by the hygrocapacity of the rock and soil. In the trap region it occurs, as a rule, scarsely and in stunted form, but there are exceptional cases in which Salai forests occurs on trap. This avoidance of trap is evident due to the fact that this area does not show sufficient retentivity of moisture. Since, on the edges of the decomposing laterite may produce a soil on which Bosewellia serrata can thrive.

PLATE 6.1

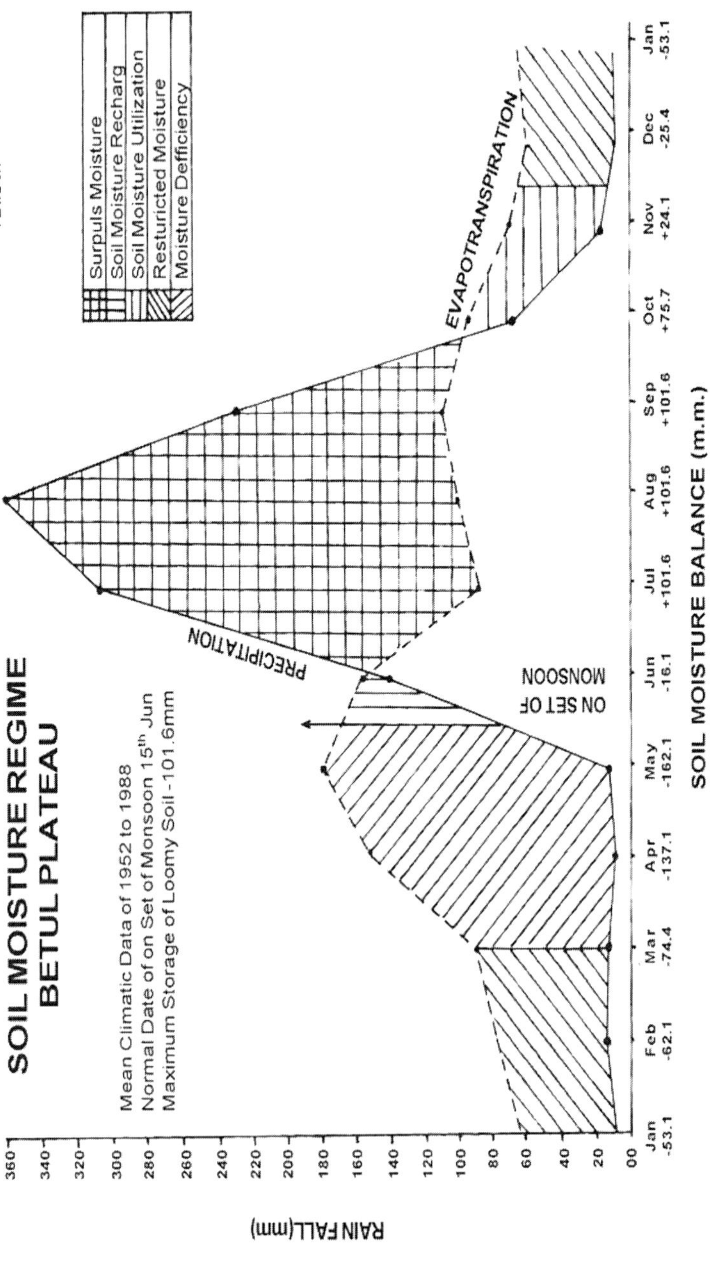
SOIL MOISTURE REGIME
BETUL PLATEAU
Mean Climatic Data of 1952 to 1988
Normal Date of on Set of Monsoon 15th Jun
Maximum Storage of Loomy Soil -101.6mm
Surpuls Moisture
Soil Moisture Recharg
Soil Moisture Utilization
Resturicted Moisture
Moisture Defficiency
PRECIPITATION
EVAPOTRANSPIRATION
ON SET OF MONSOON
RAIN FALL(mm)
360
340
320
300
280
260
240
220
200
180
160
140
120
100
80
60
40
20
00
Jan -53.1
Feb -62.1
Mar -74.4
Apr -137.1
May -162.1
Jun -16.1
Jul +101.6
Aug +101.6
Sep +101.6
Oct +75.7
Nov +24.1
Dec -25.4
Jan -53.1
SOIL MOISTURE BALANCE (m.m.)

DISTRIBUTION OF SALAI FORESTS

Though Bosewellia serrata does not exist on extensive scale in the study area, it is found scattered, all over the area. The small patches of Bosewellia serrata occur mainly on the ridges. These forests are extended over 8 per cent land part of the total forest of the region. Dry plateau areas, with dry type of forest is however, replaced by Bosewellia serrata on the southern aspect, shallow soil on rock slopes or dry tops of hills.

They are abundant in South division and cover the 6.3 per cent area of the total forest area of the division. In north and west division the Bosewellia serrata is found scarsely in association with Tectona grandis(Teak) (Table 6.1)

To the southern escarpment of the plateau, the Salai forests occur as a long belt in the ranges of the south division. The Tapti range near Betul river is covered with these forests. In rain shadow areas of the Satpura range to the northern aspect, the teak forests are replaced by the Salai forests. The Gawasen range and the Barbatpur range also consist of Salai forests. In west division, they are found as associates of teak especially on Kalibheet hill tops of Mohda and Tawadi ranges. The Salai forests are mainly confined to the area of low rainfall, shallow soil and rocky slopes and on dry tops of hills. Generally, they are of open type after leaf fall and of poor quality with low density.

In the study area, Salai forests occur in almost all forests ranges in scattered form. They are mainly concentrated in the extreme southern ranges of the region. The percentage of this forest type ranges from 0.28 in Sawaligarh forest range to 22.21 in Sawalmendha forest range. On the basis of the proportion of area under.

Salai forests, the forests ranges of this region can be classed into high, medium high, medium, medium low. The proportion of these forests is high in Sawalmendha while it is low in Saoligarh range (Plate 6.2) the range-wise distribution of Bosewellia serrata is given in Table 6.2

Table 6.1

Betul Plateau : Distribution of Salai Forests

Division	Total area under forests (in hect.)	Total area under mixed forests (in hect.)	Mixed forests area as per cent of total area	Percentage of regional area under mixed
North Division	1,23,215.80	1,015.10	00.82	08.51
South Division	1,57,078.00	9,667.90	06.15	81.10
West Division	93,112.00	1,082.00	01.16	09.08
Rampur-Bhatodi Project	30,729.50	157.50	00.51	01.31

Source: Compiled from records of the conservator of forests, Betul 1998.

Table 6.2

Betul Plateau : Range-Wise Area Under Salai Forests

No.	Range	Total Area of the Range (in hect.)	Mixed Forests Area (in hect.)	Percentage of Mixed Forest Area
1.	Saoligarh	15,351.00	109.00	00.28
2.	Amla	23,801.00	102.00	00.42
3.	Multai	23,765.00	105.00	00.44
4.	Barbatpur	25,704.10	117.40	00.45
5.	Rampur-Bhatodi	30,729.50	157.57	00.51
6.	Bhaura	29,416.50	165.20	00.56
7.	Tawdi	25,045.00	214.00	00.85
8.	Sarni	29,926.60	302.60	01.00
9.	Betul	21,659.70	221.00	01.02
10.	Gawasen	16,508.90	209.10	01.26
11.	Mohda	32,049.00	410.00	01.27
12.	Bhainsdehi	27,752.00	425.00	01.53
13.	Chicholi	20,667.00	349.00	01.68
14.	Tapti	25,441.00	667.00	02.62
15.	Athnair	27,394.00	1943.00	07.09
16.	Sawalmendha	28,925.00	6,425.95	22.21

Source: Compiled from the official records of the conservator of forests, Betul 1998.

The Sawalmendha Range

This range extends over 289 Sq. Kms. in the south of the Tapti river valley from the boundary of Maharashtra state. The Salai forest covers nearly one-fourth land of the range. In the southern part of the range crop grows as a belt from west to east. The southern slopes of the Gawaligarh hills scarping to the over looking Barar

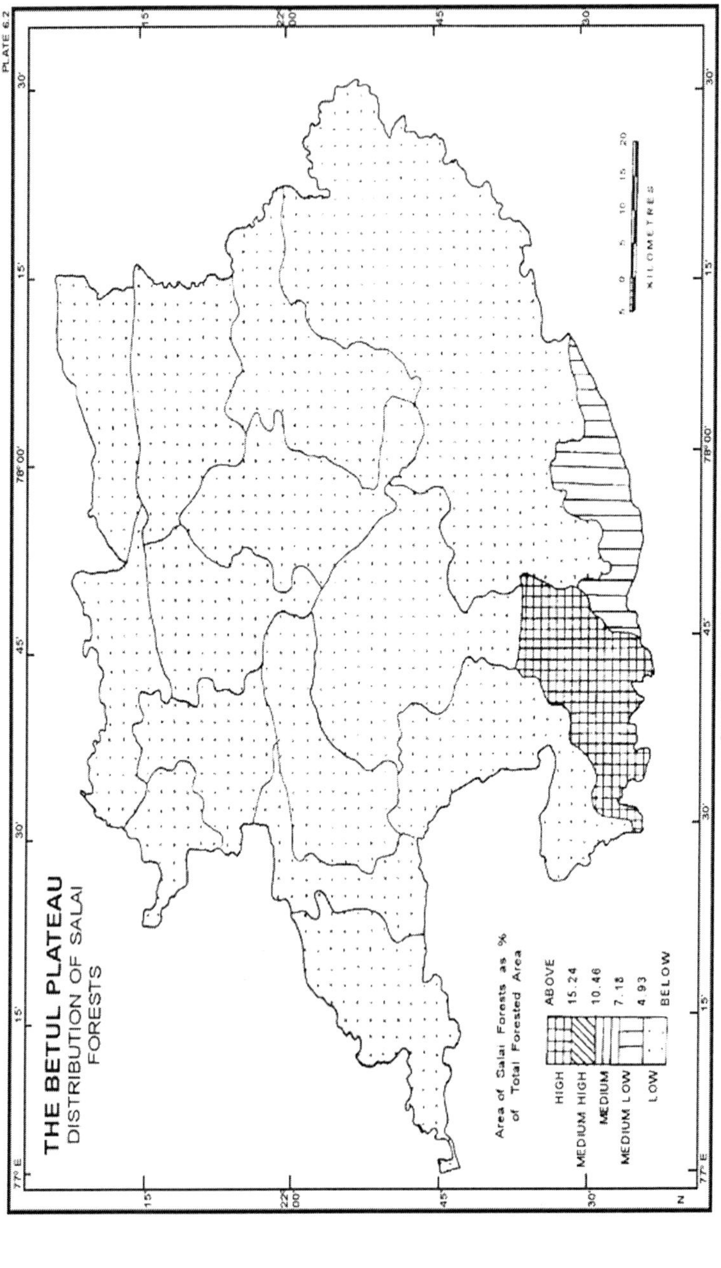

Source : Foreest Department Records

Plain are covered with extensive Salai forest due to occurrence of poor soil with stones. The Salai forest marks its presence in Patoli and Dabka forest too. The crop does not attend a large size and the density falls up to 0.5 because of erosion and mineral deficiency.

The Athnair Range

The Athnair range of south division is covered with Salai forests which occupy 194 Sq. Kms. area. In the range Salai forest are found on the southern boundary of the study area particularly in the narrow belt. The southern slope is covered with underlying rocks of the Gondwana Sandstone which has given birth to infertile, thin soil which is occupied by the Salai forest. The density varies mainly from 3 to 5.

Other Range

These forests can also be seen in western eroded hills and on rapid hilly slopes of Kalibheet hills. The soil on these hills is derived from besalt and is occupied by Salai forest. In the Central part of the study area Tapti range they are extended over the residual hills and hillocks. It is extended over higher slopes spur and escarpment parts of the north division. Thus, in these areas the forests are replaced by Salai forests due to sandy soil with scanty soil moisture.

Stratification of Salai Forests

Salai forests consist of vegetation with comparatively few species because the environment in which Salai grows, is unfavourable for the growth of other species. Though successional development of vegetation takes place in these forests but the developed species lack their parental characteristics due to varying environmental conditions. The main associates like Sterculia Urens (Kullu), Lannea grandis (Moiyan), Chloroxylon swietenia (Bhirra), Acasia catechu (Khair) Buchnania lanzan (Achar), Semecarpus anacardium (Bhilwa) grows in these forests. Advance growth of teak is often seen. The third layer is formed of shrubby species such as Holarrhera antidysentrica (Kutaja), Aloe vera (Aloe), Woodfordia frulicosa

(Dhawai), Gardenia gummifera (Dikamali), Euphorbia neriifolia (Thuhar), and Zizyphusj jujube (Ber). The ground layer consists of herbs like Acanthospermum hispidum(Gokharu),Solarium suratiense(Bhatkatiya) and Boerhaavia diffusa(Punarnava). In this layer there is always an abundance of grasses chiefly Heteropogon contortus(Kusal), Sehima sulcatum(Sainer) and Cymbapagon martini(Rosha).

VALUABLE SPECIES

Boswellia serrata/Salai Tree

It is deciduous middle-sized tree. With ash-coloured papery bark which is peeling off in thin flakes. Young shoots and leaves are pubescent. Leaves are 20-38cm. long, variables in shape, ovate or obavate lanceolate. Flowers are white at the tip of the branches. From life-forms point of view the tree falls under *Microphanerophytes* some time *Mesophanerophytes* and from leaf classes point of view the plant falls under class of *Mesophylls*.

Natural Habitat

The Salai tree is colonized in semi arid climate zones. The tract in which this tree occurs in pure form is dry and beyond the reach of monsoon rains. It shows its occurrence over hilly habitats in places where soil is poor and shallow with sheet rock cropping out. It is also found associated with Sterculia urens(Kulu) and Chloroxylon swietenia(Bhirra) on stony ground or gravelly soil. These habitats are most suitable for the existence of this tree where it shows more frequency as well as density. Its nature of distribution declined more towards the heavy soil habitat rather than that of sandy.

Phytogeographical Distribution

The distribution of the Salai tree is seen in dry deciduous forests of the study area. It is found in both, the scattered and compact form. Preponderance of Salai occurs on southern aspect, shallow soil on rocky slopes and flat hills tops. Here, it is marked in pure patches. The pure crop of the tree is also noticed in Patoli, Neelgarh and Dabka forests of Sawalmendha range and Athnair range respectively. The growth of the tree is noticed stunty in the forests of Dhabadeo in Betul range, Jamgarh in Barbatpur range, northern escarpment of Bhanwargarh hills in Bhaura range and Babri-Barra peak in Mohda range. In Asir range it is unevenly distributed.

Table 6.3

Betul Plateau : Occurrence of Boswellia serrata in the Salai Forests

Particulars	Sarni	Barbatpur	Tapti	Sawalmendha
Frequency	05.55	07.14	55.55	75.00
Relative frequency	00.16	00.32	02.14	02.66
Density	00.11	00.14	00.61	01.25
Relative density	00.18	00.27	01.60	02.84

Source: Based on field work.

Table 6.3 makes it clear that the frequency of Boswellia serrata is very high in Sawalmendha forest range, located on bordering area of southern part (75.00). The frequency of occurrence decreases from south to north direction. The lowest frequency of 5.55 is noticeable in Sarni range.

The highest density is registered in Sawalmendha (1.25) and shows a decreasing trend towards the north in Sarni (0.11). The relative density is also high in Sawalmendha but very low in Sarni range.

The wood of Boswellia serrata is used for making news print and other types of papers, so it has been cut by the forest department and supplied it as raw material to the Nepanagar Paper Mill. This over cutting has led to the destruction of tree from the forests of study area. If such ruthless over cutting is not stopped this will soon become an extinct species.

Applied Values

Nearly 19.4 per cent of the respondents have reported to be

prescribing the plant or its part as medicine in different diseases (Table 4.1). Kirtikar and Basu (1933, 521-523), Chopra and others (1956, 39), Ambasta and others (1986, 78) have also reported its use as medicine. According to them the bark of the tree is acrid and tonic which removes biliousness and allays asthma. Gum is considered to be an expectorant, diuretic and stomachic and used in diarrhoea and dysentry, pulmonary affections and cutaneous troubles. It is also useful in blood diseases, mouth sores and vaginal discharges. The fruits and flowers remove cough and cure piles. In the study area the bark of Salai is inhaled through 'Chilam' for treating gastric pain. The wood is also used for making cases, crates and boxes. It is also used as a fuel by local inhabitants. The wood may be used as a raw material for manufacturing pulp, paper and match.

Sterculia urens/Kulu Tree

It is a medium sized, soft wooded deciduous tree conspicuous in the forest by its white or greenish white shining, smooth and papery bark. Leaves are 20-30 cm. long, crowded at the ends of branches and very densely hairy on under surface. Flowers are yellowish brown, small, in very large densely hairy erect bunches. Fruits are thick and covered with stinging hairs. From life-forms point of view the tree falls in the group of *Microphanerophytes* but from leaf-classes point of view it falls under the group of '*Mesophylls*'.

Natural Habitat

The tract in which this tree occurs experiences a hot summer season and mild winter. It is found in the dry deciduous forests occurring in scattered associations, as isolated patches on higher elevation and rocky surface usually on exposed steep slopes and ridges. The open dry hilly habitat is the most suitable habitat for this plant. The tree has liking for dry shallow rocky soil. Due to its drought resistant character its development coincides in most of the habitats with the prevailing semi arid conditions. Thus, by nature it exhibits obviously the monoclimax tendency of plant succession.

Phytogeographical Distribution

The tree occurs in the tropical India. It grows wildly in Central India and Deccan plateau specially Madras Presidency.

The presence of Sterculia urens is marked throughout the hilly regions of the study area usually in Salai and dry mixed deciduous forests. It is noted upon an elevation of 820 metres in Satpura. It registers its presence in ravines of high hills, exposed slopes of hill and is also seen lonely on rocky slopes. It is the main associate species of Salai forests which is uniformly distributed on southern escarpment of Gawaligarh hills. It is abundantly found in the southern escarpment of Asir range toward Bharanga of the north eastern part of the study area. The tree also marks its occurrence in the Dabadeo and Gawasen peak of Kalibheet hills to the west. In brief, it is scarsely distributed throughout the forests of study area or it can be termed as occasional tree in the forests.

Table 6.4

Betul Plateau : Occurrence of Sterculia urens in the Salai Forests

Particulars	Sarni	Barbatpur	Tapti	Sawalmendha
Frequency	25.00	21.42	00.00	31.25
Relative frequency	00.72	00.96	00.00	01.10
Density	00.39	00.36	00.00	00.50
Relative density	00.64	00.69	00.00	01.14

Source: Based on field work.

The analytical results show that frequency of Sterculia urens is comparatively lower than the other species. Its frequency of occurrence is 31.25 per cent of quadrat in Sawalmendha, 25.00 per cent in Sarni and in Barbatpur it is 21.42 per cent. It can not be seen in the Tapti range.

The density per quadrat is registered high in Sawalmendha (0.50) and lowest in Barbatpur. But in case of relative frequency Barbatpur stand second (0.69) and highest is in Sawalmendha (1.14). In Sarni the percentage of relative frequency is (0.64)

During field observation it has been observed that the Sterculia urens is found in three fourth survey sites. It is also noted that the tree is not marked in the Tapti range because it grows only in the ecology of hilly forests.

Betul Plateau : Kullu Tree (*Sterculia urens*)

Betul Plateau : Amaltas Tree (*Cassia fistula*)

Applied Values

The medicinal uses of the tree have reported by 22.6 per cent of the total respondents (Table 4.1), Jain (1968, 166-167), Singh and Khan (1990, 33 and 201-202), Oomachan and Masih (1991, 201) have reported the medicinal properties and the uses of the plant their respective work at length. They are briefed here. The gum exudates from the stem of the tree has medicinal properties, it is called 'Karaya' gum. It is used as a mechanical laxative. It is also used as a substitute for Tragocanth gum in throat infection and in dental fixture powders, lozenges and paste. Mucilaginous extract of leaves is applied to skin to make it soft. In cases of skin diseases and other related ailments, powder of root is given with water.

Its wood for fuel purpose is not advisable for the reason that while buring the wood of Sterculia urens produces sharp gases which are harmful for eyes. This tree is suited for aforestation of stress sites in the dry regions.

Chloroxylon surietenia/ Bhirra Tree

It is middle sized tree attaining a height of 9-12 metres. Bark is corky, rough, deeply furrowed yellowish. Leaves are 15-23 cm. long and abruptly pinnate. Its flowers are branched with many flowered terminal and axillary panicles shorter than the leaves. Fruits are glabrous capsule. The tree belongs to the class of *Mesophanerophytes* and as for its leaf is concerned it belongs to the class of *Micropyhlls*.

Natural Habitat

This tree is found in dry tropical climatic zones where mean annual temperature ranges from 24°C to 30°C and average annual rainfall varies between 70cm. to 115cm. Hill tops and ridges with light and sandy soil are got occupied by Chloroxylon swietenia. The Godwana sandstones give rise to infertile soil which has been occupied mainly by Chloroxylon surietenia. Other geological formation on which Chloroxylon surietenia is typically found consisting of the part of metamorphic rocks occur in numerous localities. The tree is common on the dry foot hills often associated with Boswellia serrata and extends in some places in the regions of Anogeissus latifolia. Thus, by nature of its growth as well as for survival this tree species is characterized by polyclimax tendency of succession.

Phytogeographical Distribution

The plant is found throughout the dry deciduous forests in the Indian peninsula, extending in the north to the Satpura and Chhota Nagpur plateau.

Chloroxylon surietenia is a common associate species of the dry mixed forest and Salai forests in the area under study. It is unevenly distributed throughout the region due to habitat characteristics. In the study area Chloroxylon surietenia Dhauda sub type of forests are found on Gondwana rocks of Sarni range and in certain parts of Bhaura forest range of the north division. Chloroxylon surietenia occurs as almost pure crop on lighter soils which is sandy near Handipani forest village of Bhaura range and forest of Ghodadongri in Sarni range. In Rampur Bhatodi range, it thrives abundantly on Asir hill tops and ridges. Asir hills is the sandstone country and capped by sandy soil which favours growth of Chloroxylon surietenia. It is registered in western division on Kalibhit hills. It is frequently found on higher elevation of the Saoligarh range and Gawasen range. It is noticed on eroded high plateau and residual hillocks of the Tapti range. In the Sawalmendha, Athnair and Bhainsdehi forest range of southern division it is found abundantly on southern escarpment.

Table 6.5

Betul Plateau : Occurrence of Chloroxylon swietenia in the Salai Forests

Particulars	Sarni	Barbatpur	Tapti	Sawalmendha
Frequency	25.00	14.28	50.00	62.50
Relative frequency	00.72	00.64	01.92	02.22
Density	00.28	00.36	00.66	00.81
Relative density	00.46	00.69	01.75	01.85

Source: Based on field work.

Table 6.5 elucidates that the presence of Chloroxylon swietenia is noticed in 36 per cent quadrats. The results of the above table show that the frequency increases from north to south. The best localities of the tree are found in Sawalmendha range (62.50 per cent) followed by the Tapti range (50.00). In Barbatpur range the percentage of occurrence is the lowest but Sarni has medium frequency.

The highest density is evident in Sawalmendha (0.8) whereas the lowest is (0.28) in Sarni. Same pattern is noticed regarding relative frequency. Tapti stands second with 0.66 and 1.75 per cent density and relative density respectively.

Applied Values

The tree in useful multipurpose plant which is used commercially as fuel and as medicine.

(i) By leaving the main trunk or stem of tree, the branches after drying are used in rural areas as a fuel for the domestic purpose. Although a good quality of coal may be produced by its wood.

(ii) As timber its wood is of inferior quality, hence used for rough wood work. Thus play a vital role even in the daily life of the inhabitants.

(iii) Since the plant is of alkaline nature which contains an alkaloid of unknown constitution 'Chloroxylonine' a powerful irritant, causing dermatitis when applied to the skin. The leaves are applied to wounds. Thus it is used as indigenous medicines.

Acacia Catechu/Khair Tree

It is a moderate-sized deciduous tree attaining a height of about 15 metres or even more in some localities. There occur small paired spines on the branches. The bark is rough and it is exfoliates in long narrow strips with dark colour. Petiole is 7-10cms long with scattered prickles, pinnae are in pairs of 10-20 leaflets and are linear, imbricate, and pubescent. The leaves are shed near about

February and new leaves appear towards the end of April or during May. Flowers are pale yellow, after the appearance of new leaves and may continue till later. Fruit is a thin pod brown, shining and dehiscent containing 5-6 seeds in each pod. From life-form point of view the tree falls under *Microphanerophytes* sometime *Mesophanerophytes*. From leaf-classes point of view the plant falls under class of *Leptophylls*. Xerophytic-categorisation revealed that the tree by nature comes under the category of spiny.

Natural Habitat

Natural habitat of tree is comparatively dry regions. Due to its drought resistant character its growth as well as development coincides in dry land where the normal rainfall varies from 48 cm. to 100 cm. In its natural habitat the maximum temperature varies from 34°C to 48°C. Suited for area between 900-1200 metres above sea level.

The tree is mostly found in two type of natural habitats. Firstly, sandy alluvial beds of rivers and streams which may or may not be dry for a considerable portion of the year, here it is markably gregarious. Secondly, in dry grass lands or open forest on southern aspects, where it is less gregarious, though commonly mixed with other species.

Acacia catechu occurs on a variety of geological formations and soil, though it undoubtedly thrives best on porous, sandy and shingle and on well-drained sandstone. It occurs on granite, gneiss, schist, quartzite, basalt, limestone and laterite, while as regards soil it is common on sandy and gravelly alluvium, and on loam of gravel with varying proportions of sand and clay. It grows also on black cotton soil. It is frequent on arid shallow stony soil composed of murram or kankar.

Phytogeographical Distribution

The tree is common throughout the greater part of Indian Peninsula in dry types of mixed forest and in the sub Himalayan tract, from Jammu in the west to the eastern Himalayan foot hills and in the adjoining tracts of northeast India.

Acacia catechu is marked throughout the eroded plateau and hilly areas under study. It is scatteringly found in Salai forests and dry mixed forests. It is uniformly noticed in the forests of the Asir hill tops and its steep slopes toward the Tawa valley in the north. In Rampur Bhatodi area, it is common in dry mixed forest, often on dry stony soil associated with Bosewellia serrata, Chloroxylon swietenia and other species in the dry type mixed forests and Salai forests spreading over Gawaligarh hills to the south. In this dry region, where the rainfall varies between 80cm. to 90cm. and the soil is often poor and shallow, the trees are for the most part of small size. In the central trap plateau it occurs in open grasslands and in poor stunted type of forest where the soil has an excess of calcareous nodules on the surface. On drier and poorer ground it is stunted, but survives under conditions which are unfavourable to the existence of almost every other species. Striking examples of its hardness occur in certain parts of Chicholi plateau, on undulating ground intersected by ravines, the soil beside being very poor, is subjected to erosion. In the poorest parts of these tracts Khair occurs purely, nothing else being capable of growing but where the soil is somewhat more favourable it is associated with stunted specimen of Anogeissus latifolia, Chloroxylon swietenia, Holarrhena antidysenterica and others.

Table 6.6

Betul Plateau : Occurrence of Acacia catechu in the Salai Forests

Particulars	Sarni	Barbatpur	Tapti	Sawalmendha
Frequency	16.66	14.28	11.11	18.75
Relative frequency	00.48	00.64	00.43	00.66
Density	00.25	00.21	00.11	00.19
Relative density	00.41	00.40	00.29	00.43

Source: Based on field work.

Table 6.6 indicates that the frequency of the occurrence of Acacia catechu is not frequent and high. The highest frequency is found less than 20 per cent of the quadrats surveyed. It is highest in Sawalmendha (18.75) and lowest (11.11) in the Tapti range.

The highest density is found in Sarni range (0.25) but the highest relative density (0.43), inhabits Sawalmendha range. The lowest density and relative density is noticed in Tapti range. Hence, it can be observed that the distribution of Khair tree is uneven and scattered and rare to frequent form in the study area.

Applied Values

It is also a multipurpose plant species of Indian semi arid zone. It is used as timber, fuel, fodder commercial and medicinal plant.

1. Its sapwood is light, yellowish white, while the heartwood is deep or reddish brown. The wood is very hard and durable. The timber of Khair is used for structural purpose, as fence posts, in the constructions of houses, for making agricultural implements and also for the hubs and spokes of wheels.
2. It is also used for making furniture, railway carriages, manufacturing of pulp boards, tool-handles and other purposes.
3. The wood of this tree is a good fuel and may be converted into charcoal.
4. Wild and Domestic animals readily eat the leaves and small twigs of Khair. They are rated as good to very good fodder.
5. Katha and Cutch are the most important commercial products obtained from the heartwood of this tree.
6. Acacia catechu also have medicinal and curative properties. Its flowers mixed with milk and sugar are taken for the treatment of gonorrhoea. The Cutch and Katha have digestive properties and is a good astringent also applied externally to ulcers and eruptions on the skin *(Negi, 2003, 165)*

Buchanania lanzan/Achar Tree

Buchanania lanzan is a medium sized tree that attains a height of about 12-15 metres with straight trunk. Young branches of the tree are covered with silky hairs. Leaves are thickly coriaceous, 14-25 by 7-12 cm., broadly oblong, glabrescent above and more or less villous beneath. Shedding of the old leaves begins in November and May continue till the end of December or January and appearance of new leaves starts in March-April. Flowers are small and greenish-white. Drupes obliquely, 2 valved black, stone hard.

Under the classification of life-forms the plant comes in the group of *Mesophanerophyte*. From leaf-classes point of view it falls under the class of *Mesophylls*.

Natural Habitat

It is essentially a tree of comparatively hot and drier regions. In its natural habitat the maximum temperature varies from 38°c to 47°c and the normal rainfall from 65 cm. to 180 cm. It is found on a variety of geological formations and will grow on poor shallow soils composed of murram. It occurs nearly pure in larger or smaller patches on the laterite. In trap formation it is often remarkably gregarious, forming pure crop of greater or less extent where the trees, may grow thickly together. It is found on high plateau and foot hills up to 1500 metres. Thus, by nature of its growth as well as for survival this tree species is characterised by poly-climax tendency of succession.

Phytogeographical Distribution

It is a common species throughout the dry type forest of Central India. The distribution of species is seen throughout the study area, chiefly in dry deciduous forests of plateau areas and hilly regions. It is also found mixed up with teak forest. It is one of the associate species in the dry forests toward the west; particularly in high plateau areas and hills of Kalibheet. Here, it is found in compact patches on hill slopes of Saoligarh range, catchment areas of Labada river in Tawdi range and Babrideo hills of Mohda range. In the Central Part, its presence is marked frequently in the mixed forest belts of high plateau area which lies between Betul and Nishan River. It also registers its presence in the mixed forests of Dabka forest circle in Sawalmendha range but its growth is seen to be stunty. In the northern parts its growth is marked very well in the mixed and teak forests spreading over Satpura range. The tree is noted in scattered patches in the forests along the hill slopes and undulating and eroded plateau areas.

Table 6.7

Betul Plateau : Occurrence of Buchanania lanzan in the Salai Forests

Particulars	Sarni	Barbatpur	Tapti	Sawalmendha
Frequency	44.44	50.00	38.88	37.50
Relative frequency	01.28	02.28	01.49	01.33
Density	00.50	00.57	00.44	00.50
Relative density	00.82	01.10	01.16	01.14

Source: Based on field work.

Table 6.7 informs that the highest frequency (50 per cent) has been enumerated in Barbatpur. In all the surveyed forest ranges its occurrence is found to be occupying about 45 per cent of the quadrats. The lowest percentage of occurrence (37.50 per cent) was noticed in Sawalmendha. The highest relative frequency is seen in Barbatpur while the lowest in Sarni range (1.28).

The density per quadrat is highly concentrated in Barbatpur (0.57). Both Sarni and Sawalmendha, ranges stand second with respect to density (0.50) and lowest 0.44 in Tapti. The highest relative density is marked in Tapti (1.16) and the lowest is in Sarni (0.82) forest range. It may be stated hence, that the tree is quite widely distributed in the Salai forest belts of the study area.

Buchanania lanzan is also found in the teak and mixed forest belt of the region. Its presence is recorded in more than one fourth of the sample quadrats of the survey sites. It may be started hence, that the tree is quite widely distributed in the teak and mixed forest belts of the study area.

The tree has economic importance for the edible fruits it yields so, the fruits are collected by the local people from Chicholi, Shahpur and Betul forest range for selling them in the market. In order to get maximum yield and to avoid hard labour in gathering them, the trees are being, ruthlessly felled which is resulting in the disappearance of the species from the forests of the study area.

Applied Values

During field survey 30 persons out of 84 have intimated the use and properties of the tree. Kirtikar and Basu (1933, 659-662), Mathur (1956, 238), Pandey and Bhatnagar (1991, 211-212) have also reported the different medicinal properties and use of the Buchanania lanzan which are as described here. Its roots are acrid and remove biliousness. It also cures blood disease. Fruit is laxative and aphrodisiac cures ulcers. The oil extracted from the fruit is used as a substitute for almond oil in native medicinal preparation and confectionary. It is also applied to glandular swelling of the neck. The seeds are expectorant, stomachic and a good tonic to the body and the brain. They are useful in gleets and urinary concretions. The kernels of the fruit are used as an ointment in skin disease. It is believed to cure pimples, prickly heat and itch and also employed by women to remove spots and blemishes from the face.

Its fruits are very tasty and when ripe they become red and are locally known as Achar. They are sweet as well as sour in taste and are much liked by the native people. The fruits are collected for selling them in the market by the local inhabitants. Due to multipurpose applications kernels of the fruits have great commercial value.

Bombax Malabaricum/Semal Tree

It is a tall sized tree of about 25-30 metres height with a straight cylindrical stem and horizontally spreading branches in whorl. Its bark is grey and covered with hard sharp conical prickles. Leaves are shed at the beginning of the winter season and new leaves appear in March to April. The tree is peculiar in developing numerous flowers near the ends of the branches in the absence of leaves on the tree. Flowers are red in colour appear in January and February. The trees are usually leafless at the time of flowering. Fruits are 10-12 cm. long, capsules, woody oblong they ripen in April and May. The seeds which are numerous are surrounded by masses of white silky wool.

Natural Habitat

The Bombax malabaricum occurs in regions showing a wide range of temperature and rainfall but thrives best in the hotter regions; the maximum temperature varies from 35°C to 49°C and means annual precipitation varies from 70cm. to 200cm. or more. Generally it is found in the high plateau and on the slopes of hills. In its natural habitat, it is found up to 1000 metres in the hills. It grows in all geological formations and all kind of soil but it has liking for soil derived from trap rocks. Here it grows as a very large tree situated on shallow soil with gravelly and free subsoil at small hill.

It is often found scattered in mixed usually deciduous forests and is a characteristic tree on grassy savannah lands, where it often becomes gregarious. It grows in exposed situations on the hard trap of the Satpuras, sometimes in very dry situations. It grow on badly drained ground, but here the growth is comparatively slow and the development is poor, the trees remain stunted with tapering boles. It is often found in various unlooked for situations.

Phytogeographical Distribution

It is distributed throughout the greater part of the India except in the most arid tracts. The plant is widely distributed in the hotter parts of western and southern India. In the Dehradun valley it is one of the commonest trees on patches of shallow soil overlying hard calcareous tufa; here it is stunted. Talbot says it grows in exposed situations on the hard trap of the Akrani(Satpuras) at ,3,700 ft. altitude; the stems are often with the branches shortened and curved upwards at the ends. It is often found in various unlooked-for situations; thus it penetrates for some distance into the Himalayan valleys and may be found scattered on the slopes of the hills, sometimes in very dry situations, up to about 4,000 ft. or even more. A tree is seen at 5,000 ft. elevation above Mangoli in the Nainital hills, associated with Quercus incana. Its wide distribution is due to the fact that the cotton-covered seeds are carried by the wind to very considerable distances.

In the study area, it is found in scattered form throughout the mixed and teak forests. It is frequently noticed in the dry type mixed and Salai forests to the south but in northern part it is less frequent in the moist type mixed and teak forests. In some areas of the Athnair range toward the south with metamorphic formation, Bombax malabaricum noted in pure patches. Its presence is also marked in the moist teak forests of lower slopes of Bhanwargarh hills in Barbatpur range and moist mixed forests of Amla and Betul range. In the western part i.e. Chicholi plateau it is rare but in central part it is universely distributed on eroded high land of the Tapti range.

Table 6.8

Betul Plateau : Occurrence of Bombax malabaricum in the Salai Forests

Particulars	Sarni	Barbatpur	Tapti	Sawalmendha
Frequency	08.33	07.14	16.66	25.00
Relative frequency	00.24	00.32	00.64	00.88
Density	00.11	00.14	00.27	00.37
Relative density	00.18	00.27	00.71	00.84

Source: Based on field work.

Table 6.8 shows that the occurrence of Bombax malabaricum is registered to the particular forest ranges. It is found in about 12 per cent of the total surveyed quadrats. About 25 per cent of quadrats in Sawalmendha range enumerates highest frequency and the lowest (7.14) in Barbatpur range. Accordingly the relative frequency ranges from 0.24 per cent in Sarni to 0.88 per cent in Sawalmendha range.

The density is high in Sawalmendha and low in Sarni forest range. The medium density is marked in the Tapti range. The highest relative density is registered in Sawalmendha (0.84) and the lowest in Sarni range.

It can be observed from these accounts that the distribution of Bombax malabaricum is of scattered and sparse type. The natural regeneration of the species is being greatly influenced by the local people because they collect the ripen fruits of it for silk cotton which will result in the disappearance of the tree from the study area if serious steps are not taken.

Applied Values

The uses of this tree have been listed in the points below :

1. This beautiful ornamental tree is preferred for planting along roadside, parks and gardens.
2. Wood of this tree is soft, light and has been used more than any other wood in India for the manufacture of a cheap grade of matches owing to its abundance and accessibility.
3. The economic value of the Semal as a timber tree lies mainly in its rapid growth and volume-production. This tree may under favourable conditions yields returns higher than those yielded by trees with much more valuable timbers.
4. The capsules furnish silk wool which is used for stuffing pillows and quilts. Thus, it has also commercial importance. Local people collect the ripen fruits of it for silk wool.
5. The tree is much valued by the people for indigenous medicines. 46.4 per cent of the respondents are prescribing the plant and its parts as medicine. Dastur (1964, 37-38), Saxena and other (1990, 125), Ambasta and others (1992, 76-77) have described the drug properties of the plant in their work. The root has stimulant and tonic properties. The young dried powdered roots form the chief ingredient in the musla-semul. This medicine is aphrodisiac which is given in impotency. The bark and root are emetic. The gum has tonic and alternative properties and is used in diarrehoea, dysentery and menorrhagia. It is also used in stomach complaints. Gum is particularly useful for female disease related to their menstural disorders. Also it works wonder for the women who fails to conceive when give along with Kamarkas gond and Lodh pathani.

Woodfordia Fruticosa/Dhawai Shrub

It is a large shrub of deciduous nature with many long arching branches and numerous glandular pubescent flowers. Leaves are straggling. Bark is smooth and peeling off in fibers. Young stem is smooth and clothed with fine white pubescence. Due to more shrubby by nature it falls under the group of *Microphanerophytes* but sometime it comes under the group of *Mesophenerophytes* when it attained the size as well as shape of a tree on deep soils and on the gravel land, where its efficiency to grow is good. It may attain the full growth in the areas of good rainfall and moisture holding soils.

Natural Habitat

This shrub is found in drier parts of tropical and sub tropical regions. It is commonly found on higher elevation such as flat hill tops. In high plateau area, it registers its presence much more than low lying forest areas. The shrub is observed more in number over gravel and compact soil formations. Due to its drought resistant character its development coincides in most of the habitats with the prevailing dry conditions. Thus by nature of plant growth as well as for survival this shrub species is characterised by poly-climax tendency of succession.

Phytogeographical Distribution

The tropical and subtropical forests of India are characterised by the presence of this species. It grows mostly in dry deciduous forests of the Central India and also in parts of west and south belts of dry tropical forests.

Woodfordia fruticosa is one of the most common associate species in the miscellaneous forests of the study area. Here, it grows on undulating forest areas and in hilly areas. Its presence is marked in northern Amla range, higher slopes of Asir range, Bhanwargarh hills of Bhaura range and Harrai hills of Gawasen range. It is common in the Deccan trap plateau and trappean hills on shallow soil or physiographically dry soil area. It is frequently marked in Dabka and Patoli forests of Sawalmendha range and residual hills of Tapti range. In the Sarni and Amla range it is found on sandstone and metamorphic rocks which give birth to infertile sandy soil. Its nature of distribution declined more towards the deep soil and plain habitat.

Table 6.9 reveals that Woodfordia fruticosa grows in one third to one fifth of the quadrats. This frequency is as high as 33.33 per cent in Sarni range and as low as 22.22 per cent in Tapti range. Proportion of this plant in total plants occurring on sample site is relatively low. Dhawai plant constitutes only 0.85 per cent of the plant community in Tapti range and 1.29 per cent in Barbatpur range.

Table 6.9

Betul Plateau : Occurrence of Woodfordia fruticosa in the Salai Forest

Particulars	Sarni	Barbatpur	Tapti	Sawalmendha
Frequency	33.33	28.57	22.22	25.00
Relative frequency	00.95	01.29	00.85	00.88
Density	00.55	00.57	00.55	00.56
Relative density	00.90	01.09	01.46	01.27

Source: Based on field work.

The density is similar in all the ranges. The highest density is recordable in Barbatpur (0.57 per cent) and lowest in Sarni (0.55 per cent). The density is near about similar in both the ranges but relative density is high (1.46) in Tapti range.

Applied Values

As by nature of its usefulness for the welfare of human being, no doubt it is multipurpose shrub species of Central India. Its three applied categories :

1. Its twigs are used for fuel for domestic purpose by the local inhabitants. It produces an intense heat, so it is a good wood for fuel. Thus, wood of the plant is preferred as a fuel.
2. On undulating topography and stony habitats it is a most suitable for plantation, thus it can be used as precursor of afforestation to make a green belt.
3. The shrub is much valued by the people for local medicines. Nearly 45 respondents informed about various uses and medicinal properties of this shrub. Chopra (1956, 259), Dastur (1962, 178) have reported the uses. The dried flowers

of it is used in Haemorrhoids and is considered as a safe stimulant in pregnancy. The bark is useful in leucorrhoea, monorrhagia and toothache. It is also used in thirst erysipelas and disease of the blood. The Adiwasi of the region informed that they use the juice of flowers in case of snake bite.

Holarrhena Antidysenterica/Kutaja Shrub

It is a tall shrub or small tree, that attains a height upto 10 metres. It is a glabrous or pubescent plant species. Leaves are 10-30cm. long, ovate, thin, nerves on the leaves conspicuous. Leafstalks are very small. Flowers are white, fragrant, 1-1.5cm. in diameter, in large terminal bunches. Fruits are slender, cylindric, 20-45cm. long, 6-8mm. thick, dark grey with white specks all over. Seeds are about 1cm. long having a tuft of long (2-2.5cm.) brown hairs at top. All parts of the plant, on incision, give out white milky juice. From leaf-classes point of view it falls under the class of *Microphylls* and from life-form point of view the shrub falls under *Microphanerophytes*.

Natural Habitat

This species is found in the semi dry tracts where total annual rainfall is more than 70 cm. and in summer the mercury soars to over 40°C. Plant has occurrence over hilly and the undulating plateau up to an altitude of about 1000 metres. It is found to be very common on forest fringes in Madhya Pradesh. It has no reservation for any kind of rocks and soils. Holarrhena antidysenterica is frequently found on part of sandstone or metamorphic rock, in numerous localities. The overlying soil, as already mentioned, consists chiefly of a reddish gravelly sand or sandy loam and although the plant is often scattered among other species. Thus, by nature of its growth as well as for survival this shrub species is characterised by polyclimax tendency of succession.

Phytogeographical Distribution

The plant occurs throughout India in dry deciduous forest, chiefly in Central India. It is found to be very common on forest fringes in Madhya Pradesh and Maharashtra.

In the study area its presence is marked over dry deciduous forests. The shrub is seen most frequently in the Salai forests of flat hill top and dry plateau areas. It is also observed that the occurrence of the shrub is more common on the lower slopes of Saoligarh hills of the western part, in the belts of Salai forests of the Sawalmendha range and in Neelgarh hills of the Tapti range. On the basis of overall observation as collected from the scattered study sites over different habitats of Betul Plateau, it is concluded that the plant has occurrence mainly in two physiographic formations : firstly Graval and compact soil in high plateau and secondly, undulating hilly patches. Over such type of the habitat formation the shrub is observed with frequent to rare occurrence from distributional pattern point of view.

Table 6.10

Betul Plateau : Occurrence of Holarrhena antidysentrica in the Salai Forests

Particulars	Sarni	Barbatpur	Tapti	Sawalmendha
Frequency	16.66	28.57	44.44	37.50
Relative frequency	00.47	01.29	01.70	01.32
Density	00.27	00.50	00.55	00.50
Relative density	00.45	00.96	01.46	01.13

Source: Based on field work.

Table 6.10 elucidates that the best localities of Holarrhena antidysentrica are found in the middle parts of the region. The increasing frequency from north to the middle parts supports the above statement. The lowest frequency (16.66 per cent) is found in northern part, i.e. Sarni range while the highest percentage (44.4) of occurrence is found in Tapti range.

The density per quadrat is highly concentrated (0.55) in Tapti range and the lowest (0.27) in Sarni range. The density of the Barbatpur and Sawalmendha range is found to be similar (0.50). The lowest relative density (0.45) is registered in Sarni whereas the highest figure of it (1.46) is for Tapti range.

Applied Values

It is multipurpose economic shrub species, but three applications or uses which are as medicinal, fuel and commercial are given below :

1. Holarrhena antidysenterica is being used by 59 per cent of the persons interviewed. They informed about the different uses and medicinal properties of the plant. Sinha (1996, 101), Jain (1968, 98-99) and Chopra and others (1956, 134) have also reported its uses. It is a drug of choice for the eradication of worms in intestine. The plant bark is used in dysentery. The dried and ground bark is rubbed over the body in dropsy. Its bark is also used to strengthen the gums and lessens inflammation, excessive menstrual flow. The seeds are good in erisypelas, bleeding piles, fatigue and hallucinations. During field survey it is found that women take its seed during the time of pregnancy to normalize the original shape of stomach and the chances of any trouble during child birth are minimized. It is believed that if this drug is used with gur, the complications after delivery are averted.
2. As, it is used as indigenous medicine, its commercial importance has been enhanced.
3. Shrub exhibits the most beautiful flowers and appears very ornamental during its flowering stage, so this species is preferred for planting in parks, gardens and along roadsides.

Gardenia gummifera/Dikamali Shrub

It is a glabrous unarmed shrub. Stem consists of resinous buds. The surface is rough and fractured. Leaves are sessile or nearly so, variable in shape and size 4-7 cm. by 2-3 cm., elliptic oblong or obovate-oblong, obtuse or subacute, glabrous. The old leaves are shed during winter season and the new leaves appear in late summer. Flowers are non odorous, subsessile, 1-3 together. Corolla at first white, soon changing to yellow. Fruits are oblong with numerous longitudinal elevated lines. The plant belongs to the class of

Nanophanerophytes, from life form point of view and as for its leaf is concerned it belongs to the class of *Nanophylls*.

Natural Habitat

Its growth favours moderate climatic conditions with the mean annual temperature ranges from 24°C to 29°C and average annual rainfall is between 70 cm. to 115 cm. It is frequently seen in hilly areas and undulating plateau areas. It grows under almost all types of soil but its growth is marked very well in sandy soil. The shrub is one of the common associates of deciduous nature. Due to this one can say that the plant species by nature is charcterised by poly-climax tendency of plant succession.

Phytogeographical Distribution

The shrub is distributed mainly in tropical regions of India. It grows widely in dry deciduous forest zones of Vindhyachal, Satpura and Aravallis. In the study area, it is unevenly distributed in the dry type mixed and Salai forests. It thrives well in Salai forests of top flat hills and high plateau areas. During survey it is noticed in Sawalmendha range especially in forests of Patoli and on Neelgarh hills of Tapti range. In the teak forest areas this species is rare.

Table 6.11

Betul Plateau : Occurrence of Gardenia gummifera in the Salai Forests

Particulars	Sarni	Barbatpur	Tapti	Sawalmendha
Frequency	19.44	7.14	27.77	31.25
Relative frequency	00.55	00.32	01.06	01.10
Density	00.22	00.14	00.55	00.68
Relative density	00.36	00.27	01.46	01.54

Source: Based on field work.

Table 6.11 denotes that Gardenia gummifera is marked in all ranges but its frequency of occurrence is comparatively higher in southern and central part. It is found that 31.25 per cent of quadrats is in Sawalmendha and 27.77 per cent in Tapti range. This proportion declines to 19.44 per cent in Sarni and 7.14 per cent in Barbatpur range to the north.

The highest density (64 per hectare) is found in Sawalmendha to the south and the lowest (14) in Barbatpur to the north. In the same way the relative density is high in Sawalmendha. In the moist forest areas it is rather a more rare species. Table 5.9 shows that this shrub is commonly found in deciduous mixed forest of the study area.

Applied Values

It is indigenous medicinal plant species. Nearly 37 per cent of the total respondents interviewed prescribed this shrub as medicine. Sinha (1996, 115) and Sharma (1989, 247) have reported its Anthelmintic properties. It is a very common herbal drug available and is used as a drug of choice for the treatment of chronic pains like Arthiritis, backache and old pains of Waiste. The gum obtained from this plant is used internally in dyspepsia. The drug is anti-spasmodic, carminative, antiseptic and stimulating. It is used as Anthelmintic in cases of round worm and for cleansing foul ulcers.

ENVIRONMENTAL IMPACT ASSESSMENT

The vegetation in the Salai forests is of stable-ecology type though develops in complex environmental conditions. The influence of environmental factors can be seen clearly on-this type of vegetation. The forest ecology is limited by only some species.

The Salai forests occur in special topographic areas such as hill tops ridges and rocky slopes. in these areas it develops as a dominant species. In the area of flat dry hill tops and on the open rocky slopes they develop by xerosera process. The adjoining ridges, hills tops with light and sandy soil are occupied by Salai. Physiographically, in dry soil on all formation the occurrence of Salai is noticed.

Salai forests are generally of open type with average density of 0.5. The density may reduce still further with under stocked areas being commonly found. Mother trees of Salai having a shallow root system, by the injury to its roots during shifting cultivation, may have spread far and wide through root suckers with the possible elimination of teak and other miscellaneous species. Regeneration of Salai by seed and coppice are highly susceptible to high summer temperature and forest-fire.

The investigation shows that the study sites of Salai forests are appeared to be 'semi arid phytoclimate' and biological spectrum is appeared to be 'Therophanerophytic'.

During monsoon, grasses grow rapidly in Salai forests so the tribes of Gauli caste live there with their cattle in camps. Grazing of cattle is leading to the shrinkage of the vegetation. The Salai forests of the southern parts are found along with the Barar plains so the stealing of wood takes place across responsible for the cutting of Salai trees. They cut the trees to supply wood to Nepanagar paper industry. In this way the Salai forests are leading to the stage of disappearance. Unless these adverse factors are controlled, the growth of bushes and open Savana type of vegetation may appear to be the climax for the region.

Considering all the above mentioned factors, it may be said that the Salai forests of this area seems to be more of a climatic-climax rather than biotic climax.

CONCLUSION

The Salai forests are generally of open type vegetation with association of some species. The vegetation of Salai forests is of stable-ecology type though develops in complex environmental conditions.

In the study area, by field investigation it can be concluded that geology and disintegrated soil has played a significant role in the distribution of Salai forests. The north-east part with thin sandy soil of Gondwana system and southern most part of Gawaligarh hills with thin mantle of soil consist of Salai forest. Physiographically in dry soil on all geological formation the occurrence of Salai forest is marked. These forests occur in special topographic areas such as hill tops, ridges and rocky slopes.

Though Salai forest ecology does not exist on extensive scale in the study area it is found in scattered patches all over the area. The Salai forests cover nearly 10 per cent of the total forest area of the region. To the southern escarpment of the plateau, the Salai forests occur as a long belt in the ranges of the south division. They are confined to the areas of rain shadow of Satpura ranges.

The four distinct vegetative layers are formed by trees, shrubs, herbs and grasses respectively which are useful for the welfare of human beings in various ways, so these forests are under heavy pressure. This has increased the human activities necessitating the determination of the correct ecological status to these forests. It seems probable that these forests are biotic climax type.

REFERENCES

Ambasta, S.P., et.al., 1992, *The Useful Plants of India*, New Delhi, Publications and Information Directorate Council of Scientific & Industrial Research.

Blatter, E. and Halberg, F., 1918, The Flora of Indian Desert. J. *Bombay Nat. Hist. Soc,* 26, 218-244.

Chopra, R.N. Nayar, S.L., and Chopra, I.C., 1956, *Glossory of Indian Medicinal Plants,* New Delhi, Council of-Scientific and Industrial Research.

Dastur, J.F., 1962, *Medicinal Plants of India and Pakistan*, 2nd Ed. Bombay, D.B. Taraporevala Sons and Co. Ltd.

Guthric, J.A., Armstrong, G.R., (1961), *Western Forest Industries,* Baltimore, Joha Hopking Press.

Jain, S.K., 1968, India-The Land and the People, *Medicinal Plants*, New Delhi, National Book Trust of India.

Kirtiker, K.R. and Basu, B.D., 1933, *Indian Medicinal Plants,* Dehradun, International Book Distributors.

Mathur, R.C., 1970, *Systematic Botany,* Agra, Agra Book Store.

Negi, S.S., 2003, *Natural Resource Management in the Himalaya*, NewDelhi, A.P.H. Pub. Corporation.

Oommachan, M., and Masih, S.K., 1991, Conservation and Cultivation of Excessively Exploited and Vanishing Medicinal Plants of Madhya Pradesh, Focal Theme ISCA Symposium, Eds. Agrawal, G.P. and Hasija, S.K., Delhi, Narendra Publishing.

Pandey, R.P., and Bhatnagar, P., 1991, Wild Edible Products in Forests of Madhya, *Journal of Tropical Forestry,* July-Sept., Vol (III) 210-218.

Saxena, K.P., 1970, *A Brief Course in Economic Botany*, Lucknow, Prakashan Kendra.

Sharma, R.N., 1989, *Dehati Jadi Butian*, Delhi, Hindi Pustak Bhandar.

Singh, V.K. and Khan, Abraham, 1990, *Medicinal Plants and Folklores* (A Strategy Towards Conquest of Human Ailments), New Delhi, Today & Tomorrow's Printers and Publishers.

Sinha, Rajiv, K., 1996, *Ethnobotany, The Renaissance of traditional Herbal Medicine,* Jaipur, Ina Shree Publishers.

7

Ecological Zones

Different plant species have evolved and lived under different environmental conditions. The occurrence of plants in a region at any place depends on a number of environmental components. Therefore the special characteristics of individual, environmental factors and their relationship with the nature and distribution of forests in general and plants in particular have been discussed in the preceding chapter. The present analysis is aimed at the identification of composite ecological regions after taking into account their composite personality.

Ecological region is an area of earth surface characterised by distinct ecological responses to bio-climate as expressed by soil, vegetation, fauna etc. i.e. agroclimatic zone superimposed upto soil conditions (Mandal & Kolte).

The ecological zone is the land unit in terms which will be uniform in their bio climatic type, soil site characteristics and natural vegetation. Several attempts have been made to divide ecological zones according to different tendencies such as Blasco (1979) renamed the ten eco-floristic zones outlining the climatic features, vegetation types and floristic peculiarities of each of these. Meher-Homji (1991) delineated the phyto climatic zones mainly from the distribution of the vegetation types which themselves show close links with the climatic and edaphic factors. Subramaniyam et.al. (1984) and Ray Chowdhary et. al. (1989) demarcated agro-ecological zones of Punjab and West Bengal respectively.

These regions can be marked by using the division method of regionalization. The basic element of the ecosystem which exerts deep influence on other elements of the natural environment is geological formation. The development of landforms, soils and nature of underground water heavily depend on the geological formations. Therefore, the region has been divided into broad structural units. These geological formations have developed many typical landforms. Soils found in each of these units belong to the same broad class and thus support typical type of forests.

Along with these attributes of land, the climate goes a long way in shaping the regional flora, though the micro climate also determines the distribution, composition and the type of plant communities. The depth of data required for assessing the role of the climatic elements are lacking. Nevertheless, it is the rainfall which is the most significant climatic factor as far as the vegetative community is concerned in this tropical area. The annual rainfall of the region is 115 cm. which is taken into consideration to divide the region into two zones viz, moist zone and dry zone.

Based on the geology, landforms, soil scape and over all environmental forces of the region particularly the rainfall pertaining to the region, the plateau can be divided into ten ecological regions (Plate 7.1). Therefore present method is based on F.A.O. accepted worldwide in principle *(Sehgal et.al., 1992)* Though Geographers prefer to name the following regions as composite regions, it is the humble wish of the scholar that the regions should be known by the name ecological regions as looking to the present project which is environmental one.

1. The Dry Ecological Region of Tropics

   (i) The Asir range Ecologcal Zone

   (ii) The Saoligarh Bhanwargarh hills Ecologcal Zone

   (iii) The Deccan Trap Country Ecologcal Zone

(iv) The Tapti Basin Ecologcal Zone

(v) The Gawaligarh hills Ecologcal Zone

2. The Moist Ecological Region of tropics

(i) The Keelandeo Hills Ecologcal Zone

(ii) The Tawa Basin Ecologcal Zone

(iii) Mowar Ladi Highland Ecologcal Zone

(iv) Tapti Lowland Ecologcal Zone

(v) Khamla Upland Ecologcal Zone

THE DRY ECOLOGICAL REGION OF TROPICS

(i) The Asir Range Ecological Zone

This range extends from northeast to northwest of the Satpura system. The range is composed of sandstone and shells of Gondwana system consisting brown and grey sandstones. In the eastern part these rocks form Bijori stage comprising shells which are micaceous flages and sandstones and the western part consists rock of Motur state. The range attains the maximum height towards the east (651 metres) which gradually falls down towards the west. The Tawa river intersects this range into east and west portions. The eastern part is highly dissected by Bharanga and its upstream thus, presents a topography with steep slopes, scarpments and ridges. This region consists of sandy red or yellow soil. On the basis of climatic point of view the area is comparatively dry and hence favours the growth of dry type forests. Dense forests are also found on lower slopes and upstreams. The western part of the region has sufficient water retaining capacity due to ecological factors such as clayey shallow soil and northern aspect so this part has good quality mixed forests with Tectona grandis (Teak),Lagerstromia parviflora (Landia) and Chloroxylon swietenia (Bhirra) as dominant species. All these factors together give an identical name to the region as ecological division.

The Rampur-Bhatodi range of the study area falls in the eastern part of this section where miscellaneous species of xerophytic nature

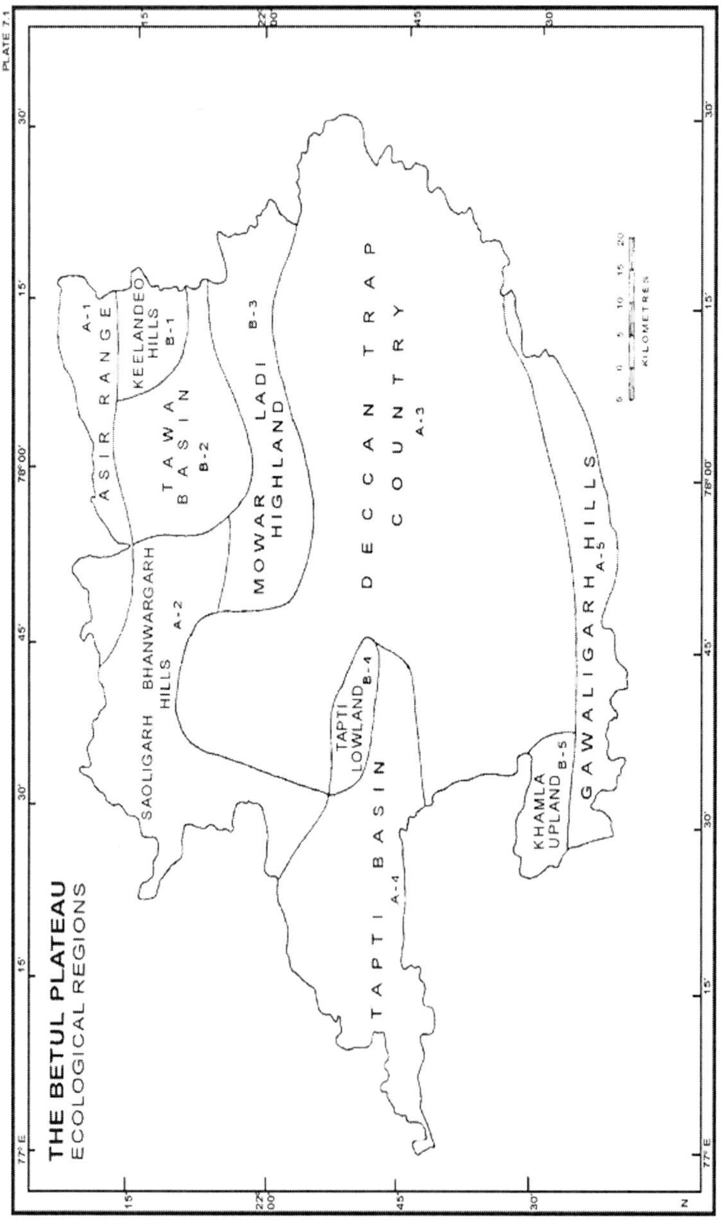
PLATE 7.1
THE BETUL PLATEAU
ECOLOGICAL REGIONS
ASIR RANGE
A-1
KEELANDEO HILLS
B-1
TAWA BASIN
B-2
MOWAR LADI HIGHLAND
B-3
DECCAN TRAP COUNTRY
A-3
SAOLIGARH BHANWARGARH HILLS
A-2
TAPTI LOWLAND
B-4
TAPTI BASIN
A-4
KHAMLA UPLAND
B-5
GAWALIGARH HILLS
A-5
KILOMETRES
77° E
78° 00'
22° 00'
N

are found, such as Acacia catechu (Kher), Sterculia urens (Kulu), Bosewellia serrata (Salai), Aegle marmelos (Bel). The western part of this section comes under Bhaura forest range where Madhuca latifolia (Mahua),Butea monosperma (Palas) trees grow frequently.

(ii) The Saoligarh Bhanwargarh Hills Ecological Zone

This ecological unit stands on north-west part of the study area. The Bhanwargarh hills stand toward the mid-northern portion of the study area whereas the Saoligarh is toward the west portion upto Tapti valley. The section is covered with Deccan trap but on hills tops, it consists of crystalline rocks. The Saoligarh hills consist of a great thick mass of hard basalt. In typical case this part shows vertical columnar joints which possesses varied topographic features i.e. the hills, scarpments, peaks and ravines. The elevation of Bhanwargarh and Saoligarh is 893 and 679 metres respectively. These hills are covered with black shallow soil. The average rainfall in this area is about 1142 mm (Shahpur). These relief features and ecological conditions have favoured a thick vegetation cover.

This ecological region includes northern Barbatpur, Gawasen, Saoligarh and Tawdi forest ranges. The microclimate in this region is humid and hence shows growth of dry deciduous type of teak forests. In this zone Tectona grandis (Teak), Terminalia tomentosa (Saj), Lagerstromia parviflora (Landia), Bamboo are the dominant species. These forests are replaced by miscellaneous species on hills where trees of medicinal value grows such as Emblica officinalis (Aonla), Madhuca latifolia (Mahua), Bombax malabaricum (Semal), Cassia fistula (Amaltas) and Woodfordia fruiticosa (Dhawai). Some valuable medicinal herbs are also noted like Asparagus adscendens (Safed musali) which is found only on Barbatpur hills of the Betul plateau.On Saoligarh hills Boerhaavia diffusa (Punarnava) and Evolvulus alsinoides (Shankhpushpi) grow abundantly.

(iii) The Decan Trap Country Ecological Zone

This country is the middle part of the plateau which is bounded by hills of Satpura range at its three sides. The trapean rocks extensively cover this part. The base consists of a thin porous layer of earthy basalt. This section is an undulating plateau area which is eroded by the Tapti, Purna, Wardha, Morand and their tributaries. It is covered with good black fertile soil derived from deccan trap. The mean annual rainfall varies between 814 mm to 1011 mm (Athnair and Multai). The region is comparatively dry but has favourable fertile soil therefore used for agricultural purpose. Normally, the region is devoid of forests, however the marginal high lands consist of dry deciduous forests. This part of the region comprises forest ranges which is known by the name of Chicholi, Athnair, Tapti, Bhainsdehi, Multai and Sawalmendha.

The ecological conditions of this country have favoured growth of teak forests, with Tectona grandis (teak) as predominant plant species. Besides some species like Azadirachta indica (Neem), Semecarpus anacardium (Bhilwa), Bombax malabaricum (Semal), Ficus religiosa (Peepal), Mangifera indica (Aam) and Butea monosperma (Palash) of great value are also found. Some other mesophytic trees found along the river beds are Terminalia arjuna (Kahu), Syzygium cumini (Jamun) and Ficus glomerata(Gular).

(iv) The Tapti basin Ecological Zone

This narrow lowland is formed by the Tapti river and its tributaries. This unit is located in the mid-western part of the Betul plateau. This basin is filled with alluvium of recent age. The ravines develop on both the sides of basin presenting an undulating surface covered with red to brownish soil. The soils of middle part are mostly derived from deccan trap. This region is drained by the Betul, Nishan. Labid, Bataki and Ambhora rivers. The average annual rainfall received by the basin is 1147 mm at Chicholi. Because of edaphic factor and microclimatic conditions the basin grows two distinctive type of forests mixed and teak.

The basin includes the forest ranges namely the Tapti, Tawadi and Mohda. In the Tapti forest range, mixed forests occupy the highest proportion (46 per cent) of the total forest area. Tectona grandis (Teak), (Saj),(Sisham) and Bamboos are the main dominant associate species. The Tectona grandis and mixed type of forests alternate each other in the sections. The trees of economic value such as Syzygium cumini (Jamun), Terminalia bellerica (Bahera), Ficus religiosa(Peepal), Madhuca latifolia (Mahua) and Azadirachta indica (Neem) are noted in this section. Some medicinal shrubs like Holarrhena antidysenterica (Kutaja),Gardenia gummifera (Dhikamali), and Cymbopogon martini (Rusa) grass grow more frequently in the basin.

(v) Gawaligarh Hills Ecological Zone

This hilly and dissected country is situated on the southern part of the study area extending from east to west. This section is covered with the deccan trap. At the southern base of the Gawaligarh hills very small exposure of archaean metamorphic rocks occur close to Salvardi. The elevation varies between 777 metres to 941 metres (Kalapakhan & Salvardi). The country receives steeply form toward the adjoining Barar plains. This region is highly dissected by the rivers of Purna and Wardha system. The ecological formations have produced major proportion of shallow black soil and red sandy soil. The northern scarpments and high peaks of these hills are covered with lateritic soil. This section receives normally 1094 mm of rainfall at Bhainsdehi. Due to varied ecological factors these hills grow distinct type of ecological dry forests i.e. Tectona grandis (teak), mixed and Salai. These hills includes 70 per cent forest area of Bhainsdehi, Sawalmendha, Athnair and Multai range where valuable plant species are found. Among them Tectona grandis, Diospyros melanoxylon (Tendu), Buchanania lanzan (Achar), Semecarpus anacardium (Bhilwa), Terminalia bellerica(Bahera),Emblica officinalis(Aonla), Sterculia urens(Kulu) and Bosewellia serrata(Salai) are the major trees. Some shrubs like Dhritkumari and Tarwid grow very well in this region. The plant species Kalimusli herb grows abundantly only in the Sawalmendha range of this section in the Betul plateau.

THE MOIST ECOLOGICAL REGION OF TROPICS

(i) The Keelandeo Hills Ecological Zone

This section is situated towards the south of Asir range in the north-eastern part of the study area. It is bounded by the river Tawa and Bharanga at its three sides. These hills have scarping edges and stands above the Tawa valley. Its maximum height is 1107 metres at Keelandeo peak. This section is covered with the rocks of lower Gondwana system of Permian era. Its northern and southern part have sandstone of Barakar and Talchir stage respectively. It consists of reddish to yellowish or brownish sandy loam with numerous intercelations of sand and gravel. The unit has micro-climatic conditions due to higher elevation thus favouring growth of moist type forests. This section includes north-eastern part of the Sarni range where mixed forests are extended. These mixed forests consist a variety of plant species. Among them some valuable plants are Terminalia chebula (Harra),Terminalia bellerica (Bahera), Aegle marmelos (Bel), Cassia fistula (Amaltas), and Emblica officinalis (Aonla).

(ii) The Tawa Basin Ecological Zone

This lowland region is located in the north-eastern part of the study area. It is bordered by hills and ranges on all sides. The north-eastern part of the section is bounded by Asir and Keelandeo hills while the Bhanwargarh hills bound its south-western part. The northern part of the basin is filled with alluvial material of recent age, whereas southern part consists of metamorphic and crystalline rocks. However, valley areas mostly consists of deep fertile soil. This section is mainly drained by the Tawa and one of its tributaries the Machna. The mean annual rainfall is 1140 mm at Shahpur. Though the northern part of this unit is put under agriculture, its major portion is covered with forests. The ecological factors such as soil and good water retentivity cause dense vegetation cover of moist type with varied flora.

About 80 per cent forest area of Sarni and 60 per cent forest area of Barbatpur range fall under this ecological region. Major portion of the basin is covered with mixed forests and rest of the area is under teak forests. Tectona grandis (Teak),Terminalia tomentosa (Saj),Adina cordifolia (Haldu), Dalbergia latifolia (Sisham),Ougeinia oojeinensis (Tinsa) and Bamboo dominant associate species are found. These mixed forests have medicinal herbs and shrubs along with large sized trees either as climbers or epiphytes. Some of them are Hemidesmus indicus (Anantmul), Plumbago zeylanica (Chitrak), Nyctanthes arbortristis (Harsingar), Tinospora cordifolia (Gilloy), Asparagus recemosus (Satavari), Euphorbia hirta(Dudhi) etc. The crop of medicinal plants is also dense and their frequency and density are very high in this region. Thus, inhabitants are much inclined towards the indigenous system of treatment and also earn money by selling them in the market for thier livelihood.

(iii) The Mowar and Ladi Upland Ecological Zone

This ecological region is located in the central part of the study area from the Machna valley to the eastern boundry of the plateau. It has Tawa basin to the north and trap country towards the south. This section is covered with granite, metamorphic and crystalline rocks of Archaean system. The granites in the Bel valley show signs of considerable disturbances. This region is a high land of Satpura plateau which is as high as 687 metres. The country is greatly undulated due to presence of residual hillocks (Matiardeo 911 metres and Paraskot 919 metres) and intersected by a large number of streams joining the Tawa. The lower slopes of the region have clayey loam while the upper parts consist of fertile sandy loam soil. This section receives 1192 mm of annual rains as recorded at Betul city. Depending on several factors like soil fertility, moisture content of soil and rainfall, the section usually have moist forests. This upland includes forests of Amla and Betul range where teak forests are found. These forests are mainly confined to the flat plateau areas which are replaced by the mixed forests at higher elevation. In the mixed forests of this region many plant species grow very well. Therefore, this unit can be said to be a suitable locality for growing different plants.This region has many collection centres for the collection of medicinal plant products like Bordehi, Amla and Betul.

(iv) The Tapti Lowland Ecological Zone

This ecological unit occupies the mid-western part of the study area and is situated towards the southern part of Chicholi. This region is composed of crystalline rocks. In fact the section is the northern part of middle Tapti basin which is drained by the river Nishan and Betul. It is an eroded area with ravines. It contains clayey loam with gravel. The mean annual rainfall is 1147 mm at Chicholi. Because of ecological conditions especially soil and microclimate, the region favours growth of moist type forests. About 35 per cent mixed forests area of the Tapti forest range falls under this section where a variety of plant species grow. Some of them are Bauhinia variegata(Kachnar), Albizzia labbeck (Siras), Scheichera oleosa (Kusum), Ficus glomerata (Gular) and Syzygium cumini(Jamun).The medicinal shrub Helicteres isora(Marorphali) grows abundantly in this region.

(v) The Khamla Upland Ecological Zone

This ecological unit is located to the southern most part of the study area known as Khamla plateau with a general elevation of 800 metres which is a part of trapean country. The Khamla peak of this upland attains a height of 1137 metres. Presence of black shallow soil is the characteristic feature of this region. Climate of this upland is comparatively cold due to its higher elevation and dense vegetation cover, hence it supports growth of moist type forests. This section occupies 16 per cent mixed forests of the Bhainsdehi forest range. Although these forests support growth of majority of the plant species, but Terminalia chebula(Harra) grows here as a pure crop. Some other trees are Bombax malabaricum(Semal), Terminalia bellerica (Bahera),Emblica officinalis(Aonla).Shrubs such as Phyllanthus niruri (Bhuiaonla), Celastrus paniculatus (Malkangini), Lantana camara (Raimunia) of commercial value are also noticed abundantly.

CONCLUSION

On the basis of above discussion it can be suggested that this region is quite suitable for growing a variety of plant species of different nature. The tree of deciduous nature such as Tectona grandis(Teak),Terminalia tomentosa(Saj), Lagerstromia parviflora

(Landia), Butea monosperma (Palas), Bamboo grow as dominant species on Trap country ecological zone. In ecological term some mesophytic tree of semi-evergreen nature like Terminalia arjuna (Koha), Syzygium cumini (Jamun), Scheichera oleosa (Kususm), Ficus glomerata (Gular), Bauhinia variegata (Kachnar), Dalbergia latifolia (Sisham) are registered in moist ecological zones Bosewellia serrata (Salai), Sterculia urens (Kulu), Acacia catechu (Kher), Aegle marmelos (Bel) trees are of Semi xerophytic nature which are found in dry ecological zones. Valuable medicinal tree Terminalia chebula (Harra) and Terminalia bellerica (Bahera) are mainly thrive in Khamla Mowar and Landi Upland ecological zones whereas Aonla is observed abundantly in Saoligarh-Bhanwargarh hills ecological zone.

In accordance with genetic characteristics the shrubs grow as mesophytes and xerophytes in the study area.Tree like shrubs such as Woodfordia fruiticosa (Dhawai), Gardenia gummifera(Dhikamali) of deciduous nature thrive in dry ecological zones, whereas some of the shrubs of evergreen nature like Millettia auriculata (Gurmar), Tinospora cordifolia (Gilloy), Hemidesmus indicus (Anantmul) and Nyctanthes arbortristis (Harsingar) are mainly confined to moist ecological zones. On dry locality xerophytic shrubs such as Aloe vera (Ghritkumari), Zizyphus jujuba(Jharberi),Zizyphus rugosa(Churna) etc are found.

The study area supports growth of a variety of herbaceous plants. These plant species are of many categories like annual, ephemeral and perennial, major proportion of which is of great medicinal and commercial value such as Asparagus adscendens (Safed Musli) and Kali Musli.

These species are probably indigenous and much more can be grown but excessive biotic activities however has accelerated the pace of destruction of useful species and day by day many of them are at the state of extinction or endangered. The propagation and preservation of these plant species is an urgent need of the day.

REFERENCES

Blasco, F., 1979, Las Territories Biographics du sous-continent Indian in Palaeogeographic et. Biographic de 1, Himalaya et du sous-continent India, Univ. Paul Sabatier Toulouse, pp. 25-30.

Mandal, C. and Kolte, K.T., Agro Ecological Zones-A Case Study of vidarbha in *Geographical Review of India,* Vol. 57, No. 4, 337-344

Mehar-Homji, V.M., 1991, Phytoclimates of Peninsular India, *Jou. of Environmental Biology,* 179 to 190.

Roy Chowdhary, C. and Mandal, D.K., 1989, Agroclimatic Zones of West Bengal, *Geog. Review of India,* Vol. 51, No. 2.

Subramaniam, A.R. and sambarive Rao, 1984, Agroclimatic Classification of Maharashtra Trans. Inst. *Indian Geographers,* Vol. 6, No. 1, Univ. of Pune.

8

Future Engineering of Forests

The study region is gifted with a wide variety of biotic resources. Management of these resources involves forest engineering. Forest engineering is meant to find out the factors which influence the growth of vegetation. It may also be defined as the application of forest business methods and principles of technical forestry to the operation of forest property *(American foresters 1958)*. Forests form an integral part of land use pattern and their management is important for delivering benefit to the people *(Rout, et. al. 1987)*. Forestry includes a wide range of technical factors from botanical, taxonomy, anatomy and physiology to engineering and from land and vegetational survey to soil sciences. It also includes mensuration of quantities and growth of trees to sociology. Forest engineering also revives the management of environmental stability, uses of rehabilitation of degraded forests, fulfillment of local requirements, attaining normality of forests for maximum forest products and also protection of wild life.

Human beings impose changes on natural eco-system. The increasing control of his environment often creates conflicts between his goals and natural processes. The present day economic system is limited upto money which circulates only within and between the ecosystem and the socio one. The explosion of human population has resulted in deforestation at an accelerated rate. Therefore, in order to manage these natural resources in general and medicinal plant resources in particular, forest engineering would be useful one.

OBJECTIVES

The main consideration of the present study is as follows :

1. Maintenance of environmental stability through preservation and restoration of the ecological balance which has been adversely disturbed by serious depletion of the forests.
2. Conservation of natural heritage by preserving the remaining natural forests with the vast variety of flora and fauna which represents the remarkable biological diversity.
3. Check on soil erosion and denudation of the catchment areas in the interest of soil and water table conservation.
4. To increase substantially the forest cover through massive afforestation and social forestry programmes especially for all denuded, degraded and unproductive land.
5. To meet the requirements of fuel wood, fodder, minor forest products medicines and timber of the rural and tribal populations by encouraging balanced utilization of forest produce and maximising substitution of wood.

The biotic factor is the biggest problem of the region particularly the man and his livestock which have burdened to forest resources to such an extent that forests are getting depleted. In the past two decades the forests cover of the region has been degraded qualitatively and quantitatively resulting into gradual deterioration in its extent and composition. Agriculture, settlements and industries exert heavy biotic pressure and faulty management practices have also led to large scale deforestation. The present ethnobotanical survey of various uses of medicinal plants though provides important information but also shows that the fast multiplying human population and their increasing demand are resulting into over exploitation of existing forest resources. It has also been observed that the indigenous medicinal plants are at an alarming state of deterioration.

These plant species are being exploited by local people and traders of course through local forest dwellers. Abundance of species have been continually on decrease. This over exploitation and some misuse of resources have been resulted into extinction of various important plant species from habitat. Therefore, there is a need to protect these existing natural plant resources in their natural habitat.

Following points also indicate the necessity of forest engineering in the study region.

(i) For the conservation of total biological diversity, the preservation and planning of disappearing wild plant species is essential.

(ii) For the long term conservation of valuable Geo-medicinal resources, it becomes necessary to apply forest engineering in the degrading forest areas.

(iii) Forest engineering also provides support to the increasing population in the area under study, as its silviculture technology will be helpful to develop Geo-medicinal resources with the participation of local people.

The concept of forest engineering started with the establishment of Central India Department in 1862. The forest officers E.E. Fernandez (1879), C.I. English (1927) and Gurdayal Singh Lamba applied various techniques of forest engineering in their working plan for the purpose of earning maximum revenue by forest products, resulting in destruction of forests from the area.

Therefore, the study area needs the operation of various Silviculture techniques in the different forest ranges for the conservation and development of plant species through forest engineering such as conversion system, selective fellings, improvement measures, coppice system and rehabilitation. Besides these some new approaches, such as social forestry, Research centre, Eco-development centres are necessary for the conservation and development of resources in the field of forest engineering. Therefore it is high time to study this aspect.

PROSPECTIVES

For long term conservation and protection and to increase productivity of regional plant resources the new technology, mechanical methodology and surveying of resources is required in the field of forest engineering. Keeping these aspects in view forest management is designed as given below : -

1. Technical Measures
2. Mechanical Measures
3. Surveying Measures

TECHNICAL MEASURES

(i) Application of Remote Sensing

Remote sensing technology will be a very patent tool in modern forestry. It is one of the techniques of scientific data collection. Colour aerial photography of this technique facilitates an easy identification of land cover, crop-condition, tree types, and tree stress and soil types.

In the study area, identification of forests and plant species, forest mapping for areas under going rapid changes can be done by remote sensing so that the life forms and phenological study of different plant species can be studied easily. It will also be a good approach to the ecological study of plants growing in different ecosystem. As it is helpful in data collection and data analysis, regarding topography, climate, surface-water, geology and soil, therefore, a suitable geo-climatic condition for plantation can be selected through this technique.

Ecological studies require detailed biotic and abiotic information which can be obtained only by ground survey, such as chemical data for the environment and organisms. However, remote sensing has provided valuable support for ground sampling methods. It is proved very useful for mapping plant associations. Already remote sensing is being used for weather analysis and forecasting seasonal changes in plant forms, gives different imageries and thus, phenology can also be studied *(Rai, 1994, 783)*. Plant structure, soil background and the surface condition of the reproductive and vegetative parts in plants are of special significance governing spectral reflections. All this can be used for studying productivity of different ecosystems which help in management & planning. Aerial photo interpretation, which is a mode of remote sensing has been used for classification of forest stands and types, survey of mortality and depletion, planning of reforestation, inventory to timber and other forest products etc.

(ii) Bio-technology

Bio-technology is the application of scientific engineering principles to the processing of material by biological agents to provide goods and services relevant for silvicultural production *(Joseph 1988, 210)*. The plants continue to remain as major source for novel photochemicals and the development of medicinal plants. Bio-technology have opened up avenues both for the propagation of elite medicinal plants and mass cultivation of cells in bioreactors for novel products *(Heble and Roja 1996, 71)*. The evolution of biotechnology in the field of silviculture, has increased the opportunities of ethnoforestry.

In the area, this technique will help to develop the valuable plant species with high and stable yield per unit area and genetic resistance to various stresses. Further yield will also be increased by optimization of nutritional treatment, control of non-crop vegetation and using advance biotechnique of tree breeding like tissue culture. The dry and eroded areas Rampur Bhatodi project and Tapti Range of the plateau can be implemented with ethnoforestry by improving tolerance capacity of plants with the help of this technology, Hence, the cultivators engaged in growing economic plants as well as foresters can be given training of bio-technology and motivated to adopt it.

Bioherbicides and biopesticides may resolve the dilemma. It will be readily appreciated that if the material needs of humans can be met by innovative use of biotechnology in the form of appreciably increasing production in an ecofriendly manner then the pressure on our planet earth will be reduced to a great extent and that will be a positive contribution, towards biodiversity conservation *(Rai 1994, 784)*. New genetic material is often necessary in order to breed strains for improved yields, nutritional quality, flavour, durability, pest and disease resistance, responsiveness to different soils and climates etc.

(iii) Non-Conventional Energy Technology

According to a study of the World Bank (Energy on Developing Countries 1980) 86 per cent of all wood is consumed in the form of fuel wood. In India woods are being removed from the forests since long ago as head-loads. Head loading engages 3 to 4 million persons in the country *(Singh and Singh 1988, 187)*. In the study area, it has been observed that about 80 per cent inhabitants use forest wood as fuel, which is leading to the degradation of significant plant species from the area, This continuous utilization of plants as fuel wood aggravate the situation. Hence, new and renewable sources of energy appear to be the most viable alternative to meet rural energy needs which are as follows:

- Bio-gas energy is the most important component of renewable energy supplies today. It forms the most convenient domestic fuel for lighting and for other energy based devices. It is non-pollutant (eco-friendly) and inexpensive energy resource and is viable alternative to meet rural energy needs.
- Solar energy is the best source of energy as it is inexhaustible abundant and can be converted directly into heat, fuel, mechanical energy and electricity. The conversion of solar energy is non-polluting and the maintenance cost of the system is negligible, Hence, in order to conserve the valuable plants from destruction as fuel, the forest dwellers can be convinced to use solar energy devices such as solar cooker, solar heater etc.
- Traditional *Chulhas* are very much energy inefficient. It has been estimated that even 1 per cent increase in the fuel efficiency of the traditional *Chulhas* may result in an annual

saving of 30 lakh tons of fuel wood. Therefore, improved *Chulhas* are good alternative which are estimated to save about 8.4 quintals of fuel wood per annum.

In this way the above mentioned tremandous non-conventional energy resources will be helpful in conserving the plants in the study area and their further destruction can be restricted by providing these energy resources i.e. bio-gas, solar energy, improved chulhas etc. to the inhabitants of rural and forest areas.

MECHANICAL MEASURES

(i) Soil Conservation

The problem of soil erosion is very serious with destruction of soil texture and disappearance of natural potential plant resources in the study area. Therefore, the implementation of the technique of effective soil conservation is highly needed. Some of the popular techniques of soil conservation are described as follows :

- Digging of staggered contour trenches about 200 per hectare in heavily eroded areas the cross section has to be increased.
- Construction of stone stabilizers on slopes.
- Plantation of soil binding vegetation like Khas, Rusha, Grass along banks.
- In the study area, agriculture is practiced on steep hill slopes under rain-fed conditions. Such slopes are better put under permanent vegetation such as fruit trees. Among the fruit trees favoured in the sub tropical hills are mango, cashewnut, chickoo and coffee shrub.
- Keelandev and Asir hills, landslides areas and Pathakheda coal mine areas need to stabilize on a priority basis in the hills. Structural measures such as construction of gabion check dams, spurs and vegetative measures such as contour wattling, mulching and planting with suitable tree and grass species.

- Attention is paid to water resource development and conservation and sediment control by programmes such as nala bunding, land shaping, water harvesting.
- The treatment of river valley (Tapti, Tawa, Purna) catchments after conservation of the reservoirs at base is best remedical measure. There is a need of preventive measures before the reservoir is filled and approach has perhaps more economic advantages as will be discussed later *(Tejwani, 1990, 290)*.

Besides these techniques, ethno forestry can also be practiced on steep slopes in place of agriculture. Such slopes can be put under permanent vegetation like trees of Sterculia urens (Kulu), Acacia catechu (Khair), Terminalia bellerica(Bahera) etc. Another way to prevent soil erosion to a great extent is, planting shrubs and bushes like Lantana camara (Raimunia), Aloe vera (Ghritkumari), Sphaeranthus indicus (Gorakhmundi) etc. across the gulley channels so that soil particles may be retained and water may flow slowly down slopes with least erosion. In this way soil may be conserved in the study area.

(ii) Waste Land Management

Current understanding equates waste land with ecologically degraded lands. These lands can be used for cultivation through proper management. Waste land management would mean making inventory of plants potentially capable to withstand inhospitable and unpredictable environment and a tentative classification of these habitat *(Bhatnagar, Shrivastava and Mishra 1990, 177)*.

The study area consists of 16 per cent waste land area of the total forest land and it is being extended gradually due to increasing demand for fuel, food, medicine etc. Therefore, any further degradation of land needs an effective halt in the process through applying waste land management techniques with tribal participation which mentioned below :

- Application of various field techniques like levelling, bunding and trenching.

- Construction of water storage tanks at the spring beds.
- Planting species of multipurpose use.

Ethnoforestry may also be developed in the study area by planting species like early colonizers, fast growers prolific seed bearers and nitrogen fixers. To improve the soil many grasses and legumes of medicinal value such as Helicteres isora (Marorphali), Cassia tora (Panwar), Mucuna prurita (Kiwanch) can be grown in waste land. Drought evading plant like Aegle marmelos (Bel), Buchanania lanzan (Achar), Madhuca latifolia (Mahua), Woodfordia fruiticosa (Dhawai), Pattharchata etc. can be grown. Likewise, the hill top can be planted with Cassia fistula (Amaltas), Sterculia urens (Kulu) and Acacia catechu (Khair) trees.

(iii) Water-shed Development

Water shed development is the rational use of land and water resources for optimum production with minimum hazard to natural forces. Soil and water conservation measures would serve as a key role in water shed management and ultimately in the recovery of vegetation. Soil measures include proper planning of land use based on land capability, such as terracing, gulley control, pasture improvement, afforestation and agricultural practice like contour farming, channel treatment etc.

On the basis of field observation and information it has been noted that high land of the study area is being degraded extensively by upland streams which are causing the disappearance of valuable plants. Therefore, propagation and conservation of these species can be done through water and soil conservation by water shed development which involves following mechanical measures :

- Construction of stop dams, brush wood check dams and gulley plug on upstreams with the help of available resources.
- Construction of unlined embankment farm pond for run off storage and recycling water for ethnoform forestry in the adjoining areas of hills and plains.
- The Gawaligarh hills lies to the southern most boundary of the plateau. In this foothill region, there are several

narrow gorges across which small earth dams can be built and water can be stored in them for subsequent use in irrigation.

- The Asir and Bhanwargarh hills and southern Mahadev range has steep hills and scarping in the Tawa valley. In these hilly forest areas construction of check dam, with size depending upon the situation, where stones are not available, earthen and vegetative check dams can be constructed.

Some vegetational measures are also helpful in watershed development like plantation of Scheichera oleosa (Kusum), Bauhinia variegata (Kanchnar) and Pongamia pinnata (Karanj) trees as well as herbs like Mentha arvensis(Podina), Mimosa pudica (Lajwanti) etc along the river beds of upstream. The technique can be installed with the help of villagers through providing them training by water shed managers. In this way villagers can also be able to identify the appropriate land use practices within the capability of land. This technique will also be helpful in rain-fed and irrigated farming, to grow medicinal crops with high value such as Ashwagandha, Asparagus adscendens (Musli) and Asparagus recemosus (Satavari).

(iv) Forest Based Industries

The forests of the study area are rich in floral medico wealth which provides basic raw material for pharmaceutical works, so the villagers of this region can set up cottage industries for the preparation of some indigenous medicines such as Rusha Oil, Triphala, Arjunarishta etc. Rusha oil is extracted from Cymbopogon martinii (Rusa) grasses that grow in Sawalmendha and Dabka range of Bhainsdehi Tehsil which may be the best place to establish oil industry. Likewise Amla and Bhainsdehi forest ranges consist of Terminalia chebula (Harra), Terminalia bellerica (Bahera) and Emblica officinalis (Aonla) trees abundantly, the fruits of which can be used to prepare an herbal drug Triphala by villagers. The Shahpur range can be selected to prepare Arjunarishta as this range has a good crop of Kahu which is the main content of the drug. Besides some more herbal drug like Chywanprash, Shankhpushpi churna, Gilloy booti, Pudinhara etc. can also be prepared in the area under study.

Above mentioned industries can be kept effective by proper planning and management with forest department in order to ensure a regular supply of raw material so that the ethno production forestry will be developed in the area.

SURVEYING MEASURES

(i) Survey of Plant Resources

The first step in the engineering of forest resources of a region is to make an exhaustive survey of all the existing indigenous plant species so that an elaborate inventory of all the medicinal plant resources can be prepared and classified. For the classification the scheme of UNESCO can be adopted which is based on function, structure, habitat and characteristics of land species. Like-wise Goldsmith's (1975) conservation course method can be applied for the evaluation of plants on the basis of object, assessment of habitat, area scarcity, number of species and vegetation structure. In this method ecological evaluation is determined in grid squares so that the endangered species can be preserved in their habitat.

(ii) Land Capability Classification of Forests

In the study area surveying measures can be implemented on the basis of land capability classification maps. In this procedure each unit of land is classified according to their capability of producing things. Soil stone ratio, soil depth, slopes, minerals found in soil and soil moisture are considered as important factors for determining land capability classification. Practice of forestry in the ecologically sensitive areas, a capability and suitability classification is very important. In the region, generalized capability cases, based on site, properties of soils, proportion of land under different slope categories, exposure to erosion hazards and dominant land use type have been identified *(Singh 1990, 116)*.

With modification of Bonner (1955) method, the first step had been to classify the land suitability for forestry. The next step was to evaluate land suitability classes - Highly suitable, moderately suitable and marginally suitable land. These classes were assessed on the basis of geomorphologically related items such as slope, soil, geology, drainage, attitude.

In the forests of the study area, land resource regions can be grouped by above mentioned procedure so that they can be planted with valuable plant species. Classification of land into land resource regions based on the available information on soils, forest types, and elevation *(Gupta et al, 1968, 34)*.

CONCLUSION

The continuous increase in population brought pressure on land, and ultimately has led to depletion of forest area. Hence the different plant species of great ecological value diminished from their native areas. Therefore, it is high time to study this aspect in its proper prospective with a view to suggest a workable model like forest management for harnessing the forest resources for optimum use.

In the study area, various programmes can be conducted to preserve the valuable plants through forest engineering. It will also help to make continuous yield demand of the fuel wood & commercial plants in the study area. For this purpose the development of forest resources is required which should be done by scientific and technical measures. Therefore, there is a need of modification in the present framework of forest management.

Various technologies like Remote-sensing, bio-technology, watershed development, wasteland management, development of non-conventional energy resources, surveying etc. should be included in forest management. In the forest area the study of plant species should be done through remote sensing in their natural habitat. Waste land management would help to cultivate the plant species of commercial value. Like wise bio-technology would play a key role towards micro propagation of elite plants.

Development of non-conventional energy resources lead to

preserve the valuable plant species which are being used as fuel by local people unknowingly. For this purpose water shed development can also be applied. As the study area is richly supplied with different plant resources, the development of small cottage industries for preparation of herbal drugs will help in utilising these valuable species. By adopting the above mentioned measures sustainable development of the region in possible.

REFERENCES

American Forestry Association, 1951, *The Progress of Forestry*, 1945 to 1950 Washington D.C.

Bhatnager, S., Shrivastava, G.P. and Mishra, R.M., 1990, An Ecological Principals of Waste Land Management, some observation for Madhya Pradesh, *Journal of Tropical Forestry,* April-June, Vol. 6(III) 177-183.

Bonnet, H.H., 1955, *Elements of Soil Conservation*, New York.

Champoin, H.G., 1936, A Prieliminary Survey of Forest types of India and burma. *Forest Research juranal,* Vol. 1, (i), p 286.

Goldsmith, F.B., 1975, The Evalution of Resources in the Country Side for Conservation Purposes, *Biological Conservation*, Vol. B, 89-96.

Gupta, S.K., Ram Babu and Tejwani, K.G., 1975, Weekly Rainfall of India for Planning Cropping Programme, *Soil Conserve, Digest,* 3 (i): 31-36.

Hoble, M.R. and Roja, G., 1996, Commercial Applications of Medicinal Plant Biotechnology in *Trends in Plant Tissue Culture and Biotechnology,* Ed. by, L.K. Pareek, Agra Botanical Publishers, 65-70.

Joseph, Von Ah, 1988, Agricultural Biotechnology and environment, A strategy for the furure, *Studies in Environment and Development*, Ed. by R.B. Singh, New Delhi Commonwealth Publishers, 209-221.

Puri, G.S., V.M. Mehu-Homji, R.K., Gupta and Puri, S., 1980, *Indian Forest Ecology,* Vol. 1, Phytography and Forest Conservation, Oxford and IBH Publishing Co., New Delhi.

Rai, A.M., 1994, Role of Emerging Technologies for Conservation of Biodiversity in *Indian Forester,* Vol. Sept, 783-784.

Singh, R.B. and Singh, M., 1988, Rural Energy in India, An Analysis of Environmental Implications, *Studies in Environment and Development*, ed. by R.B. Singh, New Delhi, Commonwealth Pub., 184-208.

Tejwani, K.G., 1990, Watershed Management in the Indian Himalaya, in *Environmental Geography*, Ed. by, R.B. Singh, New Delhi, Heritage Publishers, 290.

Bibliography

BIBLIOGRAPHY

Agarkar, S.P., 1929, Vegetation and External factors, journal of the Madras *Geographical Association,* Vol.6 (2), 29-35.

Agarwal, S.K., and Garg, R.K., 1987, *Environmental Issues and Researches in India,* Udaipur, Himanshu Publications.

Allen, S.W., 1959, *Conserving Natural Resources,* Mc. Graw Hill Book Company, Inc. New York.

Anonymous, 1983, *Research Highlights,* Central Soil and Water Conservation Research and Training Institute, Dehradun.

Anonymous, 1976, *Medicinal Plants of India,* Vol. I, ICMR, New Delhi.

Ashby, M., 1961, *Introduction Plant Ecology,* London, Macmillian.

Attiwill, P.M., and Ovington, J.D., 1968, Determination of Forest Biomass, *Forest Science,* 14, 13-15.

Barrows, H.H., 1923, Geography as Human Ecology, Anals, *Association of American Geographers,* Vol. 13, 1-14.

Barry, C., 1977, *Biogeography, An-Ecological and Evolutionary Approach,* Oxford Cox, Blackwell.

Bhatia, K.K., 1958, A Mixed Teak Forests of Central India, *Journal Ecol.,* 46, 43-63.

Biswas, S.K., 1988, *Forest Administration in India,* Allahabad, Chugh Publications.

Bose, A., 1956, Plant Life on the Vindhyan, *Nature,* 178-198.

Bose, S.C., 1948, Plant Geography of Damodar Valley, *Geographical Review,* Vol. 1 (9), 112-137.

Boyko, H., 1947, The Geo-ecological Law of Plant Distribution, *Journal Ecol.,* 335, 138-157.

Brandis, D., 1874, *The Forest Flora of North-West and Central India,* London.

Brij Gopal, 1975, *Plant Ecology and Phytogeography,* Jaipur, Rajsthan Granth Academy.

Burrough, P.A., 1986, *Principles of Geographical Information System for Land Resource Assessment,* Oxford, Oxford University Press.

Cain, S.A., 1971, *Foundations of Plant Geography,* New York, Hafner Publishing Co.

Cain, S.A., and De Oliveiro Castro, G.M., 1959, *Mannual of Vegetation Analysis,* Harpar, New York.

Champion, H.G., and Seth, S.K., 1968, *A Revised Survey of Forest Types of India Nasik,* Govt. of India Press.

Champion, H.G., 1936, A Preliminary Study of Forest Types of India and Burma, *Indian Forest Research Journal,* Vol. 1 (1).

Champion, H.G., 1920, Geology and Forest Distribution, *Indian Forester,* Vol. 46, 152-154.

Chaphekar, S.B., and Others, 1974, Ecological Studies of Monsoon Vegetation in a Crop Ecosystem in Thana District, Maharastra State, *Geobias,* 1(4), 98-100.

Chatterjee, D., 1962, Floristic Pattern of Indian Vegetation, *Proceedings of Summer School of Botany,* Held at Darjeeling in 1960, 32-42.

Choubey, K., 1986, Health Services in Madhya Pradesh, *Geographical Review of India*, June, Vol 48 (2), 49-56.

Crow, T.R., 1987, Common Regressions to Estimate Tree Biomass in Tropical Stands, *Forest Science,* Vol. 24, 110-114.

Curtis, J.T. and Cottam, G., 1959, *Plant Ecology Work Book,* Minncota, Burgess Publishing Company, 66-83.

Dakshini, K.M.M., 1971, Indian Sub-continent, Wild Land Shrubs-their Biology and Utilization, in International Symposium, Utah State University, London 3-15.

Dana, S.T., 1956, Forest Influences, *World Geography of Forest Resources,* No. 33, A.G.S. Publication.

Dansereau, P., 1957, *Bio-geography an Ecological Perspective,* New York, Ronald Press.

Dauberimire, R.F., 1947, *Plant and Environment,* New York, John Wiley and Sons Inc.

Dev, D.B., 1960, Forest Types, Study in Mainpur, *Indian Forester,* Vol.86(2), 94-111.

Drury, H., 1958, *The Useful Plants of India,* Madras, Messers Orr & Co.

Elwin, V., 1965, Field Work in Tribal World, *Folkore,* Vol. 6.

Elwin, V., 1939, *The Baiga,* London, John Murray.

Erickson, P.A., 1979, *Environmental Impact Assessment Principles and Applications,* New York, Academic Press.

Etherington, J.R., 1975, *Environment and Plant Ecology,* New York, John Wiley.

Eyre, S.R., 1968, *Vegetation and Soils,* London, Edward Amold.

Farnum, P., Timmis, R., and Kulp, J.L., 1983, Bio-technology of Forest Yield, *Science,* 219, Feb.

Forsyth, J., 1989, *The Highland of Central India,* London, Chapman and Hall.

Fuchs, S., 1960, *The Gond and Bhumia of Eastern,* Mandla, Bombay, Asia Publishing House.

Gaussen, H., Legris, P., Gupta, R.K., and Meher-Homji, Y.M., 1971, International Map of Vegetation and Environmental Conditions, ICAR, New Delhi, Institute Francais Pondichery, *Trev. Sect. Sc. Tech. Horsn seric.,* No. 17.

Gerasimov, I.P., 1983, *Geography of Ecology,* Moscow Progressive Publishers.

Gerasimov, I.P., 1975, *Man, Society and the Environment,* Moscow, Progressive Pub.

Gleason, H.A., and Cronquist, A., 1964, *The Natural Geography of Plants,* New York, Columbi University Press,

Grime, J.P., 1974, Vegetation Classification by Reference to Strategies, *Nature,* Vol. 250, 26-31.

Hagget, P., 1975, *Geography, A Modern synthesis,* New York, Harper and Row.

Hart, C.R., 1932, Forest of Madurai District, *Journal of Madras Geographical association*, Vol. XI.

Hartshorne, R., 1959, *Perspectives on the Nature of Geography,* Chicago, rand Mc. Haily.

Henry, R., Zimmer, 1948, *Hindu Medicine,* Baltimore, Johns Hopkins Press.

Hewitt, K., and Hare, K., 1973, *Man and the Environment,* Washington, D.C. Association of American Geographers, Comm. College Geography, Research Paper No. 1.

Hole, R.S., 1904, A Contribution to the Forest Flora of Jabalpur Division, Central Provinces, *The Indian Forester,* Vol. 30 (3-6) 499-514, 566-599.

Hooker, J.D., 1906, *A Sketch of the Flora of British India,* Oxford Press, London.

Jain, S.K., 1967, *Vanaspati Kosh,* A Hindi Latin Dictionary of Economic Plants of India, New Delhi.

Jain, S.K., 1963, Studies in Indian Ethnobotany II, Plants Used in Medicine by the Tribal of Madhya Pradesh, *Jammu, Bull of Regional Research Lab,* Vol. 1 (1 and 2).

Jerath, B.K., 1970, *Social and Preventive Medicine,* Ludhiana, Manmohan Jeratj.

Joseph, J., 1963, A Contribution to the Flora of Bori Reserve Forests, Hoshangabad District, Madhya Pradesh, *Bull. Bot, surv. India,* 5 (3 and 4) 281-299.

Joy, T., 1977, *Biogeography-Edinbourgh,* Oliver and Boyd.

Karnik, C.R., 1956, Flora of Tapti Valley, *Bombay Geographical Magazine.*

Karshaw, K.A., 1973, *Quantitative and Dynamic Plant Ecology,* London, Edward Amold.

Kellman, M.C., 1975, *Plant Geography,* London, Muthuen.

Khan, A.H., 1935, Forest of Anantapur District, *Journal of the Madras Geographical Association,* Vol. 9(a) 39-48.

Khoshoo, T.N., 1988, *Environmental Concerns and Strategies,* Second ed., New Delhi, Ashish Publishing House.

Kimbush, A., 1924, The Forest of South India, *Journal of the Madras Geographical Association,* Vol. 2 (I), 15-35.

Kira, T., and Shidei, T., 1967, Primary Production and Turnover of Organic Matter in Different Forest Ecosystems of the Western Pacific, *Japanese Journal of Ecology,* Vol. 17, 70-87.

Klages, K.H.W., 1942, *Ecological Crop Geography,* New York, Macmillan.

Longmen, K.A., and Jenik, J., 1974, *Tropical Forest and its Environment,* London, Longmen.

Lourenthal, D., 1972, Research in Environmental Perception and Behavior, in *Environment and Behavior,* Vol. 4, 333-335.

Maheshwari, J.K., 1963, A Contribution to the Flora of Kanha National Park, Madhya Pradesh, *Bull. Bot. Surv. India,* Vol. 5(2) 117-140.

Maheshwari, J.K., 1962, Notes on the Flora of Black Cotton Soils in East Nimar M.P., *The Indian Forester,* Feb. Vol. 88(2), 115-136.

Mani, M.S., 1974, *Ecology and Biogeography in India,* Dr. W.Junk, B.V. Publishers, The Hangue.

Mathur, H.S., 1988, *Environment Resources, The Crisis of Development,* Jaipur, R B S A Publishers.

Meher - Homji, V.M., 1962, The Bioclimates of India in Relation to the Vegetational Criteria, *Bull. Bot. Surv. India,* 4, 105-112.

Meyers, K.S., 1982, The Utilization of Forest and Other Wood Waste, *The Indian Forester,* No. 8.

Meyers, N., 1984, G.A.I.A., *An Atlas of Plant Management,* New York, Anchor Book.

Mikesell, M.W., 1974, Geography as Study of Environment, An Assessment of Some Old and New Commitments, in *Perspectives On Environment,* Commission on College Geography, Association of American Geographers, Washington, D.C., 1-23.

Mishra, R.D., 1974, *Indian Plant Ecology,* Jaipur, Rajsthan Hindi Granth Academy.

Mukherjee, S.K., 1923, Plant Succession in Satpura Range near Pachmarhi, *J. Indian Society* (3).

Narayanswamy, V., and Rao, R.S., 1969, A Contribution to Our Knowledge of the Vegetation and Flora of The Pachmarhi Plateau and Adjacent Regions, *J. Indian Bot. Society,* 39, 222-242.

Negi, S.S., 1986, *Geo-botany of India,* New Delhi, Periodical Expert Book agency.

Odum, E.P., 1979, *Fundamentals of Ecology,* Philadelphia W.B., Sounders Company.

Oomachan, M., 1977, *The Flora of Bhopal (Angiosperms),* Bhopal, S.K. Jain Brothers.

Oosting, H.J., 1956, *The Study of Plant Communities,* San Francisco, W.H. Freeman and Co.

O'Riordan, T., 1971, *Perspectives in Resource Management,* London, Pion. Ltd.

Panchin, A.J, M. Harrar, E.S.et. al., 1962, *Forest Products their Sources, Production And Utilization,* New York, Mc. Graw Hill Book Co.

Phillips, E.A., 1959, *Methods of Vegetation Study,* New York, Henery Hall and Co. Inc.

Piper, C.S., 1950, *Soil and Plant Analysis,* Adelaide, The University of Adelaide.

Polumin, N., 1960, *Introduction to Plant Geoghraphy,* London, W.I., Longman Green and Co. Ltd.

Prasad, R.R., 1987, Forests and Tribals-Issues in Policy Making, *Tribals and Forest,* edited by S.P. Sinha, Bihar Tribal Research Institute, 1-10

Puri, G.S., 1960, *Indian Forest Ecology* -A Comparative Study of Vegetation and its Environment in The India Subcontinent, New Delhi, Oxford Book and Stationary Company.

Pyle, G.F., 1979, *Applied Medical Geography,* London, Oxford Publication.

Qureshi, I.M., and Yadav, J.S.P., 1964, Forest Soil of the Arid and Semi Arid Regions of India, Symposium on Problems of Indian Arid Zone, Jodhpur.

Ram Prasad and Pandey, R.K., 1987, Study of Medicinal Wealth of Central India, *Journal of Tropical Forestry,* Oct-Dec. Vol. 3 (4), 287-297.

Rao, A.S., and Panigrahi, G., 1961, Distribution of Vegetation Types and Their Dominant Species in Eastern India, *Journal of the Indian Botanical Society,* Vol. 40.

Raphael, Z., 1945, Forest in Relation to Soil and Water, *Proc. Am Phill. Soc.,* Vol. 89.

Rashid, S.M., 1993, *Remote Sensing in Geography,* New Delhi, Manak Publications Pvt. Ltd.

Raunkiaer, C., 1934, *The Life-forms of the Plants and Statistical Plant Geography,* Oxford Clarendon Press.

Ray Chaudhuri, S.P., and et. al., 1963, *Black Soils of India,* New Delhi, National Institute of Science of India.

Razi, B.A., 1955, The Phyto-geography of the Mysore Hill Tops, *J. Mysore University.*

Ridley, H.N., 1942, Distribution Area of Indian Flora, *150th Anniversary Volume of the Royal Botanical Gardens,* Calcutta 49-52.

Sahu, T.R., 1982, The Ethnobotanical Study of Madhya Pradesh, *Journal of Ancient Science of Life,* Jan., Vol. 1 (3).

Saxena, H.O., 1971, A Contribution to the Flora of Pachmarhi (M.P.), *Bull Bot. Surv. India,* 13 (1 and 2) 79-93.

Saxena, D.P., 1976, Regional Patterns of Natural Vegetation, in *Regional Geography of Vedic India,* Kanpur, 23-30.

Seddon, B., 1972, *An Introduction to Biogeography,* Garnald Duckworth and Co. Ltd.

Shafi, M., 1950, Forests in the National Economy of India, *The Geoghrapher,* Vol. 3, No. 1.

Shah, N.C., 1982, Herbal Folkmedicines in Northern India, *Journal of Ethnopharmacol,* Vol. 6 (3), 293-301.

Shantz, H.L. and Zon, R., 1924, *Natural Vegetation,* Agric. Atlas Amer. Agric. Part-I, Sec. H., 1-29.

Sharma, D.G., 1967, The Problem of Forest Protection in India with Special Reference to Madhya Pradesh, *The Indian Forester,* Vol. 93(6), 407-410.

Sharma, P.C., and Sinha, G.N., Folklore Medicinal Plants of Ranchi (Bihar), *Nagarjun,* Vol.24 (I), 1-6.

Sharma, A.K., 1983, Cultural Evolution and Environment, *National Geographical Journal of India,* Vol. 29, 3&4, 115-120.

Sharma, S., 1980, Medicinal Plants, Love of the Tribals of Eastern Rajsthan, *Journal of Economic and Taxonomic Botany,* II, 149-155.

Sharma, P.D., 1992, *Ecology and Environment,* Meerut, Rastogi Publication.

Singh, A.K., and Doria, R.S., *Man, Development and Environment,* New Delhi, Ashish Publishing House.

Singh, R.B., 1990, *Environmental Geography,* New Delhi, Heritage Publishers.

Singh, S., 1991, *Environmental Geography,* Allahabad, Prayag Pustak Bhawan.

Singh, R.B., 1988, *Studies in Environment and Development,* New Delhi, Common-wealth Publishers.

Singh, J.S., Pandey, U., and Tiwari, A.K., 1984, Man and Forests, a Central Himalayan Case Study, AMBIO, 13(2), 80-87.

Sinha, B.N., 1958, Forest Economy of Orissa, *Indian Geographer.*

Smith, Guy and Herold, 1971, *Conservation of Natural Resources,* New York Wiley.

Stamp, L. Dudley, 1960, *Applied Geography,* London.

Stanley, A.C., 1944, *Foundation of Plant Geography,* New Delhi, Hopper and Brothers.

Stewart, J.L., and Brandi, D., 1874, *The Forest Flora of North-West and Central India,* London, William H. Allen and Co.

Stoddard, Robert H., 1982, *Field Techniques and Research Methods in Geography,* Dubuque, A, Kendall Hunt, Pub.

Strahler, A.N. et. at., 1977, *Geography and Man's Environment,* New York, John Wiley & Sons.

Subramaniam, A.R., and Rao, Sambasiva, 1984, Agro Climatic Classification of Maharashtra, *Trans. Inst. Indian Geographers,* Vol. 6, N0. I, Deptt. of Geography, Puna University.

Swarup, R., and Sikka, B.K., Development of Horticulture and Conservation of Forests in Himachal Pradesh, *Need for an Integrated Approach in Development of Hill Areas,* Ed, Papola, et.at., Bombay, Himalaya Publishing House.

Tansley, A.G., 1935, The Use and Abuse of Vegetational Concepts and Terms, *Ecology,* Vol. 16, 284-307.

Tejwani, K.G., 1981, Manpower Needs for Watershed Management in India, Proceeding at the National Seminar on Watershed Management, Dehradun, Forest Research Institute and College, 193-202.

Tiwari, P.D., and Sharma, A.M., 1989, *Tribal Ecosystem and Malnutrition in India,* New Delhi, Northern Book Centre.

Tomar, M.S. and Masleker, A.R., 1974, *Aerial Photographs in Landuse and Forest Surveys,* Dehradun, Jugal Kishore and Co.

Turrill, W.B., 1953, *Pioneer-Plant Geography,* The Phyto-geographical Researches of Sir J.D. Hooker, Netherland, Martinus Nijhoff.

Verma, D.M., Pant, P.C.C., and Hanfi, M.I., 1985, Flora of Raipur, Durg and Rajnandgaon, *Botanical Survey of India.*

Vasilyev, P.V., 1960, The Question of Geographical Study and Economic Uses of Forests, *Soviet Geography,* Vol. 1., No. 10.

Vishwanathan, M.K.,et.at., 1984, Influence of Forest on Erosion Control, *Annual Report CSWCRTI,* Dehradun.

Waheed Khan, M.A., 1962, Method of Studying Forest Types, *The Indian Forester,* Vol. 88(2), Feb. 109-114.

Watt, K.E.F., 1993, *Principles of Environmental Science,* New York, McGraw-Hill.

West, W.D., 1962, The Line of the Narmada and Son Valley, *Curr. Sci.,* Vol. 31.

Whittaker, R.H., and Woodwell, G.M., 1968, Dimensional and Production Relations of Trees and Shrubs in the Brookhavan Forest, New York, *Journal of Ecology,* Vol. 56, 1-25.

Yoda, K., 1968, A Preliminary Survey of the Forest Vegetation of Eastern Nepal, III. Plant Biomass in the Sample Plots Chosen from Different Vegetation Zones, *Journal of College of Arts and Science,* Chiba University, Natural Science Series-5, 277-302.

Doe, John B. *Conceptual Planning : A Guide to a Better Planet*, 3d ed. Reading, MA : SmithJones, 1996.

Smith, Chris. *Theory and the Art of Communications Design.* State of the University Press, 1997.

Index